DATE DUE

MAY - 8 2001	
DEC - 4 2003	
DEC 1 7 2003	

Ecologically Based Pest Management

New Solutions for a New Century

Committee on Pest and Pathogen Control through
Management of Biological Control Agents and
Enhanced Cycles and Natural Processes

BOARD ON AGRICULTURE

NATIONAL RESEARCH COUNCIL

NATIONAL ACADEMY PRESS
Washington, D.C.   1996

NATIONAL ACADEMY PRESS • 2101 Constitution Avenue, N.W. • Washington, D.C. 20418

NOTICE: The project that is the subject of this report was approved by the Governing Board of the National Research Council, whose members are drawn from the councils of the National Academy of Sciences, the National Academy of Engineering, and the Institute of Medicine. The members of the committee responsible for the report were chosen for their special competencies and with regard for appropriate balance.

This report has been reviewed by a group other than the authors according to procedures approved by a Report Review Committee consisting of members of the National Academy of Sciences, the National Academy of Engineering, and the Institute of Medicine.

This report has been prepared with funds provided by the U.S. Department of Agriculture, Animal and Plant Health Inspection Service and Agricultural Research Service, and by the U.S. Environmental Protection Agency, Office of Pesticide Policy under agreement number 59-32U4-0-28. Dissemination was supported by the W. K. Kellogg Foundation.

Library of Congress Cataloging-in-Publication Data

Ecologically based pest management : new solutions for a new century /
  Committee on Pest and Pathogen Control Through Management of
  Biological Control Agents and Enhanced Cycles and Natural Processes,
  Board on Agriculture, National Research Council.
    p. cm.
  Includes bibliographical references (p.) and index.
  ISBN 0-309-05330-7 (alk. paper)
    1. Agricultural pests—Integrated control. 2. Agricultural pests—
  Integrated control—Environmental aspects. 3. Agricultural
  ecology. I. National Research Council (U.S.). Committee on Pest
  and Pathogen Control Through Management of Biological Control Agents
  and Enhanced Cycles and Natural Processes.
  SB950.E365   1996
  632'.96—dc20   96-4946
                                                                     CIP

Any opinions, findings, conclusions, or recommendations expressed in this publicaiton are those of the author(s) and do not necessarily reflect the view of the U.S. Department of Agriculture.

Printed in the United States of America

iv

The National Academy of Sciences is a private, nonprofit, self-perpetuating society of distinguished scholars engaged in scientific and engineering research, dedicated to the furtherance of science and technology and to their use for the general welfare. Upon the authority of the charter granted to it by the Congress in 1863, the Academy has a mandate that requires it to advise the federal government on scientific and technical matters. Dr. Bruce M. Alberts is president of the National Academy of Sciences.

The National Academy of Engineering was established in 1964, under the charter of the National Academy of Sciences, as a parallel organization of outstanding engineers. It is autonomous in its administration and in the selection of its members, sharing with the National Academy of Sciences the responsibility for advising the federal government. The National Academy of Engineering also sponsors engineering programs aimed at meeting national needs, encourages education and research, and recognizes the superior achievements of engineers. Dr. Harold Liebowitz is president of the National Academy of Engineering.

The Institute of Medicine was established in 1970 by the National Academy of Sciences to secure the services of eminent members of appropriate professions in the examination of policy matters pertaining to the health of the public. The Institute acts under the responsibility given to the National Academy of Sciences by its congressional charter to be an adviser to the federal government and, upon its own initiative, to identify issues of medical care, research, and education. Dr. Kenneth I. Shine is president of the Institute of Medicine.

The National Research Council was organized by the National Academy of Sciences in 1916 to associate the broad community of science and technology with the Academy's purposes of furthering knowledge and advising the federal government. Functioning in accordance with general policies determined by the Academy, the Council has become the principal operating agency of both the National Academy of Sciences and the National Academy of Engineering in providing services to the government, the public, and the scientific and engineering communities. The Council is administered jointly by both Academies and the Institute of Medicine. Dr. Bruce M. Alberts and Dr. Harold Liebowitz are chairman and vice chairman, respectively, of the National Research Council.

Preface

At the request of the U.S. Department of Agriculture and with support from the U.S. Environmental Protection Agency, the National Research Council's Board on Agriculture convened the 14-member Committee on Pest and Pathogen Control through Management of Biological Control Agents and Enhanced Natural Cycles and Processes to assess status of the knowledge in areas of pesticide application, host resistance, and biological-control practices and to chart future direction. Specifically, the committee was charged to address the following:

- Why do we need new arthropod, weed, and pathogen control methods in crop and forest production systems?
- What can we realistically expect from investment in new technologies?
- How do we develop effective and profitable pest control systems that rely primarily on ecological processes of control?
- How should we oversee and commercialize biological control organisms and products?

Given our charge and the record of history of the application of pesticides, breeding for disease resistance, and integrating biological control practices into production agriculture, my colleagues on the committee and I deliver this report with one key message: In both science and application, researchers, providers of inputs, and growers must progress from a product based approach to an *ecologically based pest management system* identified as EBPM. Management is the key word. In fact, the word *control*, as in biological control, is misleading. Pests in most cases cannot be controlled; pests must be managed with the objectives of a safe, profitable, and durable outcome.

With a better understanding of ecology, the inherent strengths of the managed ecosystem can be used with more modest inputs than in the past. Essentially, the change to EBPM as proposed here will require a substantial change from the primary practice of product input to the primary mind set of information and management. Ultimately, EBPM will help to address ecosystem health not by administering products alone to treat symptoms, but by integrating components that maximize use of natural processes with minimum development of resistance.

EBPM will require regulatory oversight that matches the level of risk of biological inputs added to the managed ecosystem. For example, synthetic chemicals are new to the biosphere—they have no base of performance in the environment or in relation to human health. However, biologically based organisms, products, and resistant cultivars are inherently different, for the most part, from synthetics. Biological processes, having existed in nature over time, provide a base of experience that is a major resource to evaluate the safe application and establish appropriate oversight of EBPM. Biologically based products are not inherently different from synthetics in their vulnerability to development of resistance, although history suggests that such will be less frequent. Users will need to monitor managed ecosystems for early identification of pest resistance.

In this report we place major emphasis on the research information needs and on appropriate regulatory oversight. The committee also urges an interactive, cooperative approach to development of EBPM. Given the other individuals and organizations addressing issues relating to the adoption of new pest management approaches, we have only modestly considered adoption in our report.

In this deliberative report the Executive Summary presents the findings and key recommendations. Chapter 1 describes the history of pest management and the limitations of current practices. Chapter 2 details the committee's new approach to pest management, ecologically based pest management. Chapter 3 identifies priority research areas and discusses important institutional changes to effectively carry out that research. Chapter 4 assesses regulatory oversight and aspects of risk assessment and management.

The contents of this report offer a new paradigm, the concept of EBPM. We are optimistic that the development and application of the principles of EBPM will contribute to a future with high-quality food, fiber, and forest production and sound management of our natural resources for safety, profitability, and durability.

RALPH W. F. HARDY, *Chair*
Committee on Pest and Pathogen Control
through Management of Biological
Control Agents and Enhanced Natural
Cycles and Processes

THE COMMITTEE ACKNOWLEDGES with deep appreciation all those who contributed their expertise to this project. The committee is especially grateful for the contributions of Joseph Panetta of Mycogen Corporation. Mr. Panetta provided useful ideas and insights based on his experience with registration and commercialization of biological control products.

Contents

PREFACE   vii

EXECUTIVE SUMMARY   1
  Cultural and Biological Approaches to Pest Management   2
    Safety, Profitability, and Durability   3
  Developing a Knowledge Base   4
    Understanding the Interactive Processes of Ecosystems   5
    The Need for Multidisciplinary Ecosystem Research 6
    New Research Methods   6
    Implementation Research   7
    Information Inputs   7
  Public Oversight of Ecologically Based Pest Management   8
    Appropriate Risks   8
    Experience and Experimentation   9
    The Need for Guidelines   10
  A New Era   10

1 LESSONS FROM THE PAST PROVIDE DIRECTION
  FOR THE FUTURE   11
  A Brief History of Pest Management Using Naturally
    Occurring Substances   12
    Early Biological Management of Arthropods   12
    Early Biological Management of Weeds   13
    Early Biological Management of Diseases   17
  A Brief History of Cultural Practices   17
  Plant Breeding   21

Synthetic Organic Pesticides 23
    Insecticides 24
    Herbicides 24
    Declining Use 24
Integrated Pest Management 25
Obstacles to Continued Use of Broad-Spectrum Pesticides 26
    Problems and Limitations of Pesticides 26
    Problems that Defy Conventional Chemical Solutions 29
    Human and Environmental Health Concerns 37
Time to Reassess and Plan 41

2 DEFINING AND IMPLEMENTING ECOLOGICALLY BASED PEST MANAGEMENT 42
Goals of Ecologically Based Pest Management 42
Supplements to Natural Processes 44
    Biological-Control Organisms 46
    Biological-Control Products 47
    Synthetic Chemicals 47
    Resistant Plants 47
Economic Feasibility of Ecologically Based Pest Management 49
    Economic Feasibility of Pest Management 49
    Economic Feasibility and Risk 54
The Role of Information in Pest Management 56
Role of Collective Action in Pest Management 63
    Grower Cooperatives 63
    Small-Market Support 64
    Certification 64
    Monitoring Pests 64

3 ACCELERATING RESEARCH AND DEVELOPMENT 69
Foundations of a Knowledge Base 70
Priority Research Areas 71
    Research on the Ecology of Managed Ecosystems 72
    Research on Behavioral, Physiological, and Molecular Mechanisms to Effect EBPM 76
    Research to Identify and Conserve Natural Resources Needed for EBPM 82
    Development of Better Research and Diagnostic Techniques 84
    Development of Ecologically Based Crop Protection Strategies 86
    Research on Implementation and Evaluation of EBPM 88
    Research to Improve Understanding of the Socioeconomic Issues Affecting Adoption 89
    Development of New Institutional Approaches to Encourage the Necessary Interdisciplinary Cooperation 91
Infrastructure for Research 94

4 PUBLIC OVERSIGHT OF ECOLOGICALLY BASED
    PEST MANAGEMENT 95
  Human Health Risks 98
  Environmental Risks 100
    Nontarget Effects 100
    Exacerbation of Plant Pests 105
  Risk Assessment and Management 108
    Drawing on Experience and Experimentation 108
    Setting Priorities 109
    Managing Risk 111
    Gaps and Inconsistencies in Current Oversight 112
    Options for Improvement 114

REFERENCES 117

ABOUT THE AUTHORS 133

INDEX 137

Executive Summary

Pests—arthropods, weeds, and pathogens—have been, are, and will continue to be major constraints to agricultural production and forestry in the United States and throughout the world. Synthetic chemical pesticides were introduced in the 1940s and used widely on agricultural crops in the hope that they would control agricultural pests. It is now clear that their use has some unfortunate consequences. Many consumers believe that trace residues of synthetic chemicals in food are undesirable and represent a significant food safety risk. In some cases, undesirable environmental impacts of synthetic pesticides have caused consumers to oppose the use of these materials in agriculture and caused governments to regulate or outlaw their use.

Pests develop resistance to synthetic chemical pesticides, just as microbial pathogens of humans develop resistance to antibiotics. In fact, pest resistance currently limits the efficacy of many insecticides, fungicides, and herbicides; and there are pests such as plant-parasitic nematodes and bacteria for which no effective pesticides are available. Many synthetic chemical pesticides are broad-spectrum, killing not only arthropod and pathogen pests but also beneficial organisms that serve as natural pest-control systems. Without benefit of the natural controls that keep pest populations in check, growers become increasingly dependent on chemical pesticides to which pests may eventually develop resistance. Thus there is an urgent need for an alternative approach to pest management that can complement and partially replace current chemically based pest-management practices.

Pest-management strategies can be viewed in context of whole-farming systems. In whole-farming systems, pest-management methods are integrated into

1

other management components of agronomic systems such as crop fertilization, cultivation, cropping patterns, and farm economics.   Such alternative farm-management strategies that promote soil and plant health, and water quality were recommended by the National Research Council in their report *Alternative Agriculture* (National Research Council, 1989b).

CULTURAL AND BIOLOGICAL APPROACHES
TO PEST MANAGEMENT

Ecologically based pest management (EBPM) is recommended as a profitable, safe, and durable approach to controlling pests in managed ecosystems.

Early agriculturalists faced serious pests that decimated crops. Through trial and error, they implemented practices that challenged such pests. Successful strategies predominantly were those that served to maintain the ecological balance of the region and the natural balance of agricultural pests and their enemies. This committee believes that such practices, combined with the advanced biological technologies now available, are the most logical approach to developing a profitable, safe, and durable (long-lasting and self-maintaining) approach to pest management. The systems, hereafter identified as ecologically based pest management (EBPM), as outlined by this committee, rely primarily on inputs of pest biological knowledge and secondarily on physical, chemical, and biological supplements for pest management. The EBPM systems will be built on an underlying knowledge of the managed ecosystem, including the natural processes that suppress pest populations. It is based on the recognition that many standard agricultural practices disrupt natural processes that suppress pests. In contrast to standard practices that disrupt and destabilize the agroecosystem, agricultural practices recommended by EBPM will augment natural processes. These practices will be supplemented by biological-control organisms and products, resistant plants, and narrow-spectrum pesticides.

The concept of EBPM builds on the cultural and biological approaches to pest management that were in use prior to the widespread application of synthetic chemical pesticides. Many practices, such as crop rotation, fallowing, intercropping, and incorporation of organic matter into soils, served to conserve and foster populations and activities of biological control agents that were indigenous components of traditional agricultural ecosystems; nevertheless, pest outbreaks and crop and animal disease epidemics did occur. Through trial and error, early agriculturalists discovered and used natural substances, beneficial organisms, and selected resistant plants to their benefit. These early agriculturalists laid the foundation for pest management based on the biology of the ecosystem. The EBPM system builds on these earlier approaches but uses advanced technological tools and methodologies to improve the knowledge base and, where necessary, includes inputs—chemical, biological, or physical—that meet the objectives of safety, durability, and profitability.

The objectives of EBPM are the safe, profitable, and durable management of pests that plague agricultural production and forestry. Examples of EBPM systems currently in use are included in this report to demonstrate the potential of EBPM to meet the goals of safety, profitability, and durability, but it is now necessary to move EBPM beyond isolated instances and into the mainstream of pest management. Wide-scale implementation of EBPM will require generating a substantial ecological knowledge base of agricultural and forest ecosystems and development of regulations and oversight consistent with the risks of the inputs. The process of creating the necessary knowledge base must begin now because the information needed to form the foundation of EBPM is lacking. Knowledge must then be transferred effectively to growers and producers to enable the implementation of EBPM. The need to generate the necessary knowledge and to develop ways to facilitate its transfer, in addition to the need to develop appropriate regulations, are emphasized in this report because these are viewed as major factors that currently limit implementation of EBPM.

Early integrated pest management (IPM) innovators should be recognized for conceiving a framework for an integrated approach to arthropod, weed, and pathogen pest management. IPM strives to manage pests using ecological principles of natural pest mortality factors; pest-predator relationships; genetic resistance; and cultural practices (National Research Council, 1989b). This theoretical basis of IPM is similar to EBPM.

The practice of IPM, unfortunately, is not always consistent with the theory of IPM. The focus of early IPM programs was devoted to control of insects, setting a precedent for the focus of IPM on arthropod pest management. In many cases, this management was limited to pest scouting and precise applications of insecticides. Consequently, this focus has been at the expense of IPM of weed and pathogen pests.

The ecological concepts of the IPM framework are the departure point for this new information-rich management strategy. EBPM will rely on an improved knowledge base of the complex ecological processes that occur in plant production.

Safety, Profitability, and Durability

The three fundamental goals of EBPM are (1) safety, (2) profitability, and (3) durability. The committee considers all three of these goals essential to developing and implementing EBPM strategies.

EBPM is a total systems approach designed to have minimal adverse effects on nontarget species and the environment because biological controls are limited by their specificity with respect to the target pest organism or by their distribution or persistence in the environment. In addition, the acute and chronic toxicities associated with conventional chemical pesticides generally have not been found with biological-control organisms and products, although there can be risks such as allergenic reactions with multiple exposure. However, unlike new synthetic chemicals, whose effects on living organisms and the environment may be difficult to predict, with biological organisms and molecules, there is in most cases an experience base on which environmental risks can be assessed.

For EBPM to be successful, it must also be profitable for the grower. Growers demand safe, economical, and effective tools that provide long-term management of pests; and these needs must be met before growers will implement EBPM's new tools. Because producers will implement only those pest control methods that lower economic risks and enhance profits, they will insist on assurances that biologically based tools are cost-effective and provide consistent responses. Alternative management strategies may indeed be less expensive than chemically based methods, but information to determine relative costs is not available and needs to be researched.

EBPM strategies also must be long-lasting to have a positive impact on crop protection. Work by crop breeders to increase cultivar resistance to arthropods and pathogens is constantly being undone by pests that overcome the plant's resistance. As new pest-resistant cultivars are introduced into a cropping system, biotypes of organisms that overcome this host resistance can predominate. Similarly, pesticide resistance limits the durable use of chemical pesticides; and resistance of pests to biological-control products and organisms can also occur. Decreasing the pest's rate of development of resistance to new inputs is essential for the durability of EBPM.

Given the funding to build the necessary knowledge base, scientists can further develop ways to modify the deployment of biological-control organisms, products, and resistant cultivars to delay the onset of resistance. A large annual research and development investment of more than $500 million continues to support the production of organic pesticides; an increased investment in research of biological processes will be needed to support the development of EBPM.

DEVELOPING A KNOWLEDGE BASE

A national research agenda, both general and cross-cutting, should be developed to identify broad areas of ecological research that promise to yield the critical information needed to accelerate progress in EBPM.

Implementation of EBPM will require more ecological and economic information than do current pest-management systems. With conventional pest-management systems, growers rely primarily on synthetic chemical inputs to control a broad spectrum of pests. With EBPM, growers and pest managers depend on knowledge of pest biology and economic feasibility of pest control options. Because numerous organisms can cause crop damage and reduce yields, growers need information about many pest-management alternatives to make an informed decision.

Understanding the Interactive Processes of Ecosystems

Because pest management should be based on knowledge of agricultural and forest ecosystems, researchers should focus on developing a better knowledge base of the interacting components and processes that characterize agricultural and forest ecosystems.

An agricultural or forest ecosystem consists of a dynamic web of relationships among crop plants or trees, herbivores, predators, disease organisms, weeds, etc. These organisms constantly evolve and respond to each other, creating a diverse, complex, and ever-changing environment. The occurrence of a pest in a particular agricultural field is not an isolated event and, as such, must be studied in relation to the larger geographic region.

EBPM uses knowledge of interactions among pests and naturally occurring beneficial organisms to modify cropping and forestry systems in ways that reduce rather than eliminate damage caused by pests; thus, EBPM is designed to enhance the inherent ecological strengths of the system. External inputs—a continuum of biological, physical, and chemical tools—would be added only if they promote the long-term environmental health of soil biota, crops, and other organisms of agricultural and forestry production systems. In contrast to the current approach, which often relies predominantly on broad-spectrum chemical methods, with EBPM, new inputs—natural or synthetic—must target specific pests and minimize disruptions to the managed ecosystem.

Research is needed to identify organisms and their functions in the ecosystem. Little is known about the diverse competitors, predators, and parasitoids that reside in soil and plant environments; yet, beneficial organisms that control pests within their native environments have been found in these habitats. Researchers are discovering that organisms in a cropping system interact in many ways—through competition, molecular signaling, toxicity, host selection, predation, and antibiosis. Understanding the processes of these interactions can lead to science playing a decisive role in controlling pest populations and contributing to the stability of natural systems. Based on these discoveries, researchers are developing new pest-management strategies to hold pest populations in check.

EBPM requires development of biological-control organisms that can be used to mimic natural processes of pest suppression. Identification of biological-control organisms is leading to commercialized products that can effectively suppress specific pests with increased safety to human health and the environment. Biologically based supplements will include live biological-control organisms (for example, beneficial insects, nematodes, microbes, and viruses), the application of specific biological-control products (pheromone attractants and microbial toxins), and development of genetically improved plants (disease-resistant cultivars). Both traditional and molecular methods will be used for genetic improvement of crop plants and beneficial organisms.

The Need for Multidisciplinary Ecosystem Research

The complexity of managed ecosystems necessitates coordinated multi-disciplinary and interdisciplinary research to develop and implement EBPM.

The knowledge needed to build the foundation of EBPM will come from nearly all disciplines of both the biological and social sciences and will require research from molecular to landscape levels. The committee has identified eight broad research areas that should receive priority:

1. ecological research of managed ecosystems,
2. research of behavioral, physiological, and molecular mechanisms affecting EBPM,
3. identification and conservation of resources necessary for EBPM,
4. development of better research and diagnostic techniques,
5. development of ecologically based crop protection strategies,
6. research on implementation of EBPM,
7. research on socioeconomic issues of EBPM, and
8. development of new institutional approaches to encourage interdisciplinary cooperation.

Although individual research will continue to be important in developing the information and conceptual framework for EBPM, interdisciplinary research and cross-disciplinary flow of information will be needed. Institutional barriers that fail to foster interactive endeavors will continue to impede research and training needed to build the fundamental knowledge necessary to develop EBPM. A professional society that unifies scientists involved in all aspects of agricultural research, development, and implementation can lead efforts to achieve EBPM.

New Research Methods

Researchers should devise new methods to study, monitor, and evaluate agricultural and forestry ecosystem processes and to develop effective pest-management tools.

New methods must be developed to study, monitor, and evaluate pests and potential biological-control organisms in order to expedite the acquisition of knowledge needed to implement EBPM. The lack of effective methodologies to characterize ecological systems hampers research aimed at generating the conceptual framework of agricultural and forestry ecosystems on which EBPM will be based. The development of molecular techniques useful in genetic manipulation of plants, insects, and microorganisms provide an unprecedented opportunity for optimization of host plant resistance or biological-control activity; but these techniques must be amended substantially to expand their usefulness from model organisms to the diverse groups of plants and biological-control organisms that will comprise EBPM. Researchers, pest control advisors, and growers will require sophisticated methods for tracking and monitoring populations of pests and biological-control organisms to optimize timing of applications of biological or chemical supplements and to avoid unnecessary risks associated with the application of supraoptimal concentrations of these supplements. New site-specific techniques such as geographic information systems (GIS) and remote sensing already are useful in geographic applications and can potentially be useful in tracking and monitoring pest movements.

Implementation Research

Implementation research is needed to facilitate the transfer of technology from researchers to pest managers in the field. Experience with IPM indicates that the complexity of EBPM may become a major limitation to its implementation. Laboratory discoveries must be moved into agricultural practice to increase adoption of EBPM. Pest suppression activities of biological-control organisms can be demonstrated in the field to growers and pest managers, emphasizing the site-specific nature of EBPM methods. Growers need analyses that compare benefits and costs of pest-management methods before they can make informed decisions. Scientists must devise strategies to supply growers with sufficient biological-control inputs and conserve valuable biological-control resources, thus increasing the longevity of EBPM. Implementation of EBPM will be advanced only if information is shared among scientists, suppliers of agricultural products, pest managers, and growers.

Information Inputs

In the past, growers relied primarily on cooperative extension agents to provide pest-management education. The public sector should continue to play an important role in training pest-management personnel and consultants. With increasing knowledge and time requirements needed to implement EBPM strategies, information will need to be synthesized to increase its accessibility. Independent consultants will play a more important role in the transfer and synthesis of information, and the increase in automated information sources will facilitate information transfer at all levels.

Transfer, synthesis, and simplification of information among growers, suppliers, and public agencies will speed the successful development and implementation of EBPM; and public investment to develop and approve registration of biological-control organisms can assure availability of safe and effective pest-management tools. Extension scientists, in particular, need to become active and interested participants to make EBPM successful. Cooperative activity among growers, suppliers of agricultural supplements, and investors as well as the scientific community would lead more quickly to breakthroughs such as discoveries of ways to delay development of pest resistance and, thus, reduce pest-management costs. Such cooperative efforts among all necessary communities will benefit stakeholders and society at large.

PUBLIC OVERSIGHT OF ECOLOGICALLY
BASED PEST MANAGEMENT

Wide-scale implementation of EBPM could require thousands of commercialized biological-control organisms, products, and resistant cultivars, each of which could be quite specific with respect to the target pest as well as the cropping system to which each could be applied. Public oversight is required to ensure that potential risks to human health or the environment are properly assessed and managed, thereby promoting public acceptance of the use of biological-control organisms and products or resistant cultivars. In assessing risk, public agencies must use appropriate criteria and methods to avoid delays and expedite reviews. In the same way that public funds are used to obtain registration of minor use pesticides, there is a need to encourage the registration of biological-control organisms with limited market potential.

Appropriate Risks

Evaluation of risks associated with deployment of biological-control organisms and products and resistant plants should be based on evidence relative to both the type of organism or product and its method of deployment.

Humans will come into contact with biological-control organisms, biological-control products, and resistant cultivars during production and application and through exposure to organisms or residues that persist on crops and in the environment. An advantage of the tools of EBPM is that many have negligible toxicity to humans; nevertheless, their potential risks to human health must be assessed. Large-scale production of biological-control organisms and products should minimize human exposure to bacteria, fungal spores, proteins, and other reactive materials. Resistant cultivars, altered by classical breeding or genetic engineering, may affect human health if greater amounts of toxic metabolites are produced in the edible portion of the plant. Similarly, biological-control products and metabolites produced by biological-control organisms should be evaluated for human toxicity.

Much has been learned from prior releases of biological-control organisms; this information can provide a basis for assessment of future risks associated with the use of biological-control organisms in agriculture and forestry. Biological control has a credible history; in the many years in which biological-control organisms and resistant plants have been used in agriculture, negative effects on nontarget organisms or on plant health have been observed in very few cases. If a biological-control organism is unfamiliar or there is uncertainty about the environment into which it is introduced, however, careful evaluation must precede its introduction. Nontarget organisms closely related to the pest are most at risk from parasitism or predation by introduced natural enemies, although quantification of such risk requires more evaluation. There is also a small possibility of exacerbating pest problems following introduction of biological-control organisms or resistant plants. For example, progeny resulting from genetic exchange between a resistant plant and indigenous plant species could persist and could possibly become troublesome, invasive components of agricultural ecosystems. Although the long history of deployment of resistant plants indicates that such problems occur extremely rarely, the risk of environmental effects resulting from the use of resistant plants in agriculture must be assessed. Unknown consequences of genetic exchange between biological-control agents and indigenous organisms also may exist, but field evaluation can assess the likelihood and effects of these events in a natural setting.

Experience and Experimentation

All experience including that gained from the natural occurrence of the biological materials and their toxicity and field testing should be fully considered in regulatory review of new organisms or products.

Knowledge of their origin in the biosphere provides an appropriate experience base for most biological organisms and molecules. For all EBPM candidate organisms or products, this experience base should be combined with experimentally derived data as well as data about organisms that are closely related taxonomically or functionally as evidence to assess risk. The evidence should be organized so as to address the potential risk criteria specific to the organism, product, or plant and its deployment. Experience with these manifestations of risk will be useful in evaluating EBPM strategies.

The potential exists for managing pests in an agricultural or forestry ecosystem using a wide range of biological controls; this, however, creates the necessity of setting oversight requirements for each biological product or organism. The task of risk assessment can be expedited by drawing on experiences with similar biological-control organisms or products, pests, and agricultural and forest ecologies; and as this knowledge base expands, oversight requirements can be adjusted accordingly.

Experience, experimentation, and expert opinion should direct oversight attention to broad-spectrum organisms, products, or resistant plants and uses on major acreage crops where risk impact could be greatest. At the same time, effective review will exempt or remove from oversight those organisms, products, or resistant plants for which accumulated experience indicates low risk.

The Need for Guidelines

The U.S. Environmental Protection Agency and the U.S. Department of Agriculture, both currently responsible for oversight or pesticide regulations, should develop and publish a guide to risk criteria, data requirements, and oversight procedures that apply to importation, movement, introduction, testing, and release or registration of biological-control organisms or products.

Inconsistencies in the existing laws and regulations are barriers to effective, efficient oversight of biological-control organisms, products, and resistant plants. Complexities and anomalies of the current regulatory system may be attributed to the overlapping jurisdiction of several agencies, the diversity of organisms to be regulated, and the attempt to make the decision-making "template" developed for registration of conventional chemical pesticides applicable to biological controls. It is essential that regulatory agencies assess risk using criteria and protocols that are appropriate to biological tools in contrast to broad-spectrum synthetic chemical pesticides. This will provide an appropriate level of oversight and minimize costs imposed by duplication of oversight and reporting requirements.

A NEW ERA

This committee believes that it is necessary to refocus objectives from pest control to pest management based on maintaining natural ecological balances. The problems of consumer and societal acceptance, safety, pest resistance, and economic cost dictate the need for immediate change in current pest-management practices that rely on short-term, broad spectrum solutions to solve major pest problems. There needs to be a paradigm shift in pest-management theory from managing components or individual organisms to an approach that examines processes, flows, and relationships among organisms. Major barriers must be overcome if EBPM is to be successful. It cannot be overemphasized that collaboration of scientists in a range of disciplines, suppliers of agricultural products, educators, and growers is critical to the success of EBPM. Many believe that the consultant/farmer linkage will be where knowledge, decision making, and action take place. The lack of such collaboration contributed to a retreat from the principles of IPM. This committee sees an opportunity to move beyond IPM and into an information-rich era in which collaborative efforts break down current barriers among the disciplines, institutions, and philosophies to achieve ecologically based pest-management solutions that are safe, profitable, and durable.

1

Lessons from the Past Provide Direction for the Future

Since the practice of agriculture began, humans have been struggling to reduce the adverse effects of pests on crops, forests, and other human-managed ecosystems. In crop production, the development of diverse cropping systems, breeding and selection for pest-resistant plants, and beneficial cultural practices helped boost yields and reduce losses caused by pests. The same principles, applied to forests and other managed ecosystems, also helped to protect them against pest damage.

Agriculture, however, was still vulnerable to periodic crop failures and pest destruction, such as the 1845 potato blight in Ireland that led to famine and widespread malnutrition. Technological advances, such as the development of Bordeaux mixture in 1883 to control fungal diseases, helped lessen crop losses. The synthetic, organochlorine and organophosphate insecticides, the herbicide 2,4-D, and the halogenated hydrocarbon fumigants introduced after World War II revolutionized pest management in agriculture. Additional synthetic chemical pesticides including fungicides, nematicides, and other herbicides and insecticides were developed and put to widespread use.

Initially, the benefits these new chemicals brought to agricultural production were thought to be without major disadvantages; however, ecological and human health risks and the economic costs of heavy, widespread use of broad-spectrum chemical pesticides are becoming more apparent. In addition to the potential for adverse effects on human and environmental health, there are growing concerns about the durability of current approaches to pest management. The disruption of inherent natural and biological processes of pest management, the resistance to pesticides developed by many major pests, and the frequency of pesticide-induced or -exacerbated pest problems suggest that dependence on pesticides as the dominant means of controlling pests is not a durable solution. The failure to develop economically viable pesticides for some of the most damaging pests and the economic costs of continual pesticide application has also led to an interest in alternative approaches to crop protection.

Managing the pests inherent to agroecosystems is imperative to producing adequate supplies of food and fiber to meet current and future world needs. Effective, long-lasting, ecologically sound, and affordable pest management systems are essential to agricultural productivity and profitability. The committee concludes that agriculturalists must learn more about the biological and ecological processes of the crop-production environment in order to develop the management approaches and products that alone or in combination with carefully managed use of selective pesticides will provide novel and lasting solutions to pest problems.

A BRIEF HISTORY OF PEST MANAGEMENT USING NATURALLY OCCURRING SUBSTANCES

The earliest known mention of using naturally occurring compounds to manage pests was in 1000 B.C. when Homer referred to the use of sulfur compounds. In the western hemisphere, early agriculturalists reduced the number of arthropod pests by treating infested plants with tobacco extracts and nicotine smoke. The list of naturally occurring substances used as pesticides expanded over time to include rotenone, soaps, fish oil, lime-sulfur, and copper-sulfate. A shift toward chemical controls coincided with the use of Paris green (an arsenic compound) in 1867 to control an outbreak of Colorado potato beetle in the United States and the fortuitous discovery of a fungicide mix (copper sulfate and hydrated lime) in 1882 in Bordeaux, France (Bottrell, 1979).

Early Biological Management of Arthropods

The use of beneficial organisms to manage pests also has a long history. The Chinese introduced colonies of ants into citrus groves to control caterpillars as early as 324 B.C. In 1752, Linnaeus wrote about the use of predatory arthropods to control arthropod pests on crops. An early insectary design in which caterpillars were placed in cages to attract beneficial arthropods was recommended by Thomas Hartig of Germany in 1827. In one of the first attempts at classical biological control, entomologist Asa Fitch suggested that toadflax, an exotic plant introduced to the United States from Europe, could be managed by importing the natural enemy of this weed from its native habitat. The colonization of the cabbage butterfly in the U.S. mid-Atlantic region and parts of the Midwest compelled Riley and Bignell to import a parasitoid (*Apanteles glomeratus*) in 1883 that eventually spread and successfully controlled the caterpillar pest throughout the United States (DeBach and Rosen, 1991).

These efforts set the stage for the most dramatic and well-known early application of a beneficial organism in the United States—the introduction of the Australian ladybird beetle (*Rodolia cardinalis*) to control the cottony-cushion scale (*Icerya purchasi*) in California. Throughout the 1880s, the cottony-cushion scale devastated the California citrus industry; by 1890, all scale infestations were under control following release of the ladybird beetle. This success was credited with saving the California citrus industry and catalyzed the expansion of biological control of other pests, including other arthropods, weeds, and diseases (DeBach and Rosen, 1991).

Early Biological Management of Weeds

In the 1930s the prickly pear cactus (*Opuntia* spp.) was successfully controlled by a weed-feeding scale (*Dactylopius opuntiae*) on Santa Cruz Island off the coast of California (Goeden, 1993; Turner, 1992). One of the most notable achievements, however, occurred on rangeland where devastation by Klamath weed (*Hypericum perforatum*) so decreased land values that ranchers were not allowed to use their ranches as collateral to borrow money for chemicals to control the noxious weed. In 1946 the arthropods *Chrysolina hyperici* and *C. quadrigemina* were released and by 1956 had consumed 99 percent of the Klamath weed, increasing the price of land 300 to 400 percent (DeBach and Rosen, 1991).

The first organized attempt at biological control of an aquatic weed was directed against alligator weed (*Alternanthera philoxeroides*). More than 30 arthropods that parasitize this weed were found in South America in the weed's native range. Of these organisms, a flea beetle (*Agasicles*), was found to be safe and potentially effective and was introduced into the United States. Today, alligator weed remains under check in most of the South; however, the effectiveness of the flea beetle has been reduced in Mississippi as a result of extensive aerial spraying of chemical insecticides, and the flea beetles alone are usually not sufficient to manage this weed in high visibility areas such as golf course ponds and irrigation canals.

For water hyacinth, plant pathogens were suggested as possible biocontrol organisms as early as the 1930s in India. It was not until 1970, however, that a serious commitment was made by the University of Florida, the U.S. Army Corps of Engineers, and Florida Department of Natural Resources to explore the potential of microbial pathogens to control aquatic weeds. Several pathogens of alligator weed, water hyacinth, hydrilla, and Eurasian water milfoil were soon discovered, and three fungi—*Cercospora rodmanii, Fusarium culmorum,* and *Mycoleptodiscus terrestris*—were studied in detail for development as bioherbicides for water hyacinth, hydrilla, and Eurasian water milfoil, respectively, in Florida, Mississippi, and Massachusetts (Charudattan, 1990a).

Selected History of Pest Management

Discovery	Date	Implementation
	B.C.	
	1000	Homer refers to sulfur use in fumigation and other forms of pest control.
	324	Chinese introduce ants (*Acephali amaragina*) in citrus trees to control caterpillars and large boring beetles.
	A.D.	
	70	Pliny the Elder notes pest control methods from Greek literature in the preceding 3 centuries.
First published observation of arthropod parasitism.	1602	
	1669	Earliest mention of arsenic used as insecticide in Western world.
	1690	Tobacco extracts used as contact insecticide.
	1752	Linnaeus recommends use of predatory arthropods to control arthropod pests.
	1821	In England, sulfur used as fungicide on mildew.
	1845	Italian Society for Promotion of Arts and Crafts awards gold medal to Antonio Villa for successful use of arthropod predators to control arthropod pests.
	1858	Pyrethrum first used in United States.
Mendel publishes paper proposing existence of "hereditary factors."	1866	
Millardet discovers value of the Bordeaux mixture.	1883	
	1889	Australian ladybird beetle introduced to control cottony-cushion scale; credited with saving the California citrus industry.
TEPP, first organophosphate insecticide discovered.	1938	*B. thuringiensis* first used as microbial insecticide.
	1938	Breeders release wheat cultivar resistant to stem-rust (*Puccinia graminis*), a fungal disease.
In Switzerland, DDT discovered to be insecticidal.	1939	2,4-D synthesized as an analog to plant hormone indole acetic acid.

Selected History—continued

Discovery	Date	Implementation
	1942	First DDT shipped to United States for experimental use; introduction of 2,4-D, first of hormone (phenoxy) herbicides.
	1943	Dithiocarbamate fungicide introduced.
	1944	Fumigant 1,2-dichloropropane (D-D) developed.
	1945	Chlordane, first persistent, chlorinated cyclodiene insecticides, and first carbamate herbicides introduced.
	1946	First house flies resistant to DDT observed in Sweden.
	1948	First U.S. registration of microbial pesticide (*Bacillus popilliae* + *B. lentimorhus*) to control Japanese beetle larvae.
First synthesis of a synthetic pyrethroid, allethrin.	1949	Captan, first of dicarboximide fungicides, developed.
Watson and Crick discover the double-helix structure of DNA.	1953	
	1954	Release of the spotted alfalfa aphid-resistant cultivar.
	1958	Atrazine, first triazine herbicide, and paraquat, first bipyridylium herbicide, introduced.
Stern et al. publish treatise on pest management; it becomes the foundation for IPM.	1959	
Nurenberg deciphers the genetic code.	1960	*Bacillus thuringiensis* "Berliner" registered to control lepidopteran larvae.
	1962	Carson's *Silent Spring* published.
	1969	Arizona places moratorium on use of DDT.
	1970	U.S. Environmental Protection Agency (EPA) established.
	1971	Cohen and Boyer develop recombinant DNA technology using restriction enzymes.
	1972	EPA cancels nearly all uses of DDT.
	1974	EPA sets first standards for worker reentry into pesticide-treated fields based on dermal toxicity of pesticide.

continued

Selected History—continued

Discovery	Date	Implementation
	1975	EPA cancels all nontermiticide uses of aldrin and dieldrin; registration of first virus (Heliothis nuclear polyhedrosis) for budworm/bollworm control on cotton; first arthropod growth regulator (methoprene) registered.
	1977	First registration of a pheromone (gossyplure for pink bollworm) for use on cotton.
	1979	*Agrobacterium radiobacter* registered to control crown gall disease.
	1980	Protozoan *Nosema locustae* registered to control grasshoppers.
	1981	Mycoherbicide DeVine®, using the fungus *Phytophthora palmivora*, registered to control citrus strangler vine.
	1983	First successful transfer of a plant gene from one species to another.
	1986	Development of transgenic virus-resistant plants using coat protein gene.
	1987	First U.S. field trials of transgenic plants (tomatoes with a gene for arthropod or virus resistance).
	1988	*Bacillus thuringiensis* "San Diego" and *B. thuringiensis* "tenebrionis" registered to control coleopteran larvae.
	1990	Fungus *Gliocladium virens* registered to control *Pythium* and *Rhizoctonia*.
	1994	Regulatory approval of transgenic virus-tolerant squash, arthropod-tolerant cotton, herbicide-tolerant soybeans and cotton

Early Biological Management of Diseases

Early in this century, plant pathologists recognized that (a) both native and exotic microorganisms could suppress plant diseases; (b) disease suppression could be manipulated through the use of cultural and management practices that altered soil's organic matter, temperature, or pH; and (c) disease suppression was attributed to the presence of suppressive microorganisms (Cook and Baker, 1983). The first attempts to suppress plant diseases by introducing beneficial microorganisms to soil were done in the 1920s (Cook and Baker, 1983), and the fungus *Phlebia gigantea*, the first biological-control organism used commercially for control of an aerial plant disease, was described in the 1950s. In addition, inoculation of cut pine stumps with *P. gigantea* prevents infection of the stumps by the pathogenic fungus *Heterobasidion annosum* and its subsequent spread to neighboring standing trees through root grafts. This biological control is still widely used today in commercial pine plantations in England and Sweden. Inoculation of chestnut trees with hypovirulent isolates of the chestnut blight pathogen *Cryphonectria parasitica* was initiated in France in the 1960s, where it successfully suppressed chestnut blight in many regions. Biological control of crown gall with the bacterium *Agrobacterium radiobacter* strain K84, which was initiated in the 1970s, is now used to suppress the disease in orchards and nursery stock worldwide. Today, at least 30 different biological-control organisms are available as commercial formulations for suppression of plant diseases (Lumsden et al., 1995).

As early as 1922 the United Fruit Company identified land types in Central America in which banana plants were termed either short-lived or long-lived crops. Both were grown in soil containing wilt pathogens, but only the long-lived plants survived because they were planted in soil that suppressed the pathogen *Fusarium oxysporum* (Cook, 1990). Since then, this natural process of suppression of fusarium wilt has been demonstrated for carnation, cucumber, cotton, flax, muskmelon, and tomato crops.

In the 1930s, knowledge of medicinal antibiotics was applied to plant pathogens by Weindling, a plant pathologist, who researched the use of *Gliocladium* and *Trichoderma* fungi for their antibiotic effects on damping-off disease. In 1938 Linford and colleagues reported on the use of a nematode-parasitic fungus to control root-knot nematodes (Cook, 1990). The knowledge gained by these applications of biological controls continues to be used today in managing plant pathogens, weeds, and arthropods.

A BRIEF HISTORY OF CULTURAL PRACTICES

Early agriculturalists realized the benefits of cultural practices such as rotating crops, altering planting and harvesting schedules, planting mixtures of crops, managing irrigation and drainage, and removal of crop residues, to reduce pest

Experimental Demonstration for Importance of Natural Processes

Researchers have used pesticides to experimentally eliminate biological-control organisms of pest species as a way to quantify the importance of biological and natural processes in managing pests. The resulting pest infestations and crop damage can be used as an estimate of the degree to which biological and natural processes keep pests in check.

California Red Scale (*Aonidiella aurantii*) on Citrus

Experimental DDT applications caused a pest population to increase from 36-fold to more than 1,200-fold over a period of several years (DeBach and Rosen, 1991).

Cottony-Cushion Scale (*Icerya purchasi*)

Two experimental applications of DDT to trees with initial light-to-moderate scale populations initially produced scale infestations sufficient for tree defoliation or death within 1 year. With lighter infestations, four treatments over a 2-year period were required to achieve the same level of infestation (DeBach and Rosen, 1991).

Citrus Red Mite (*Panoncychus citri*)

Two and one-half months after one application of DDT, the mite population index on treated trees was 2,303 compared to 377 on untreated trees. "The use of DDT on citrus in California was abandoned early in the game by citrus growers because of its obvious effect in causing such mite increases" (DeBach and Rosen, 1991:p. 11).

Cyst Nematode

Application of soil fungicides to research plots in England revealed the widespread phenomenon of biological control of cereal cyst nematode by naturally occurring soil fungi: populations of cyst nematodes remained low in untreated plots

growth and reproduction. Many of these cultural practices were effective because they augmented natural processes of pest suppression. Natural processes include predation, parasitism, pathogenesis, competition, and production of antibiotics by organisms that coexist with pests. In their native environments, plant pests live in communities composed of a variety of organisms, including their natural enemies (i.e., biological-control organisms), which constrain the pest populations and activities by natural processes. Through trial and error, early agriculturalists, building on natural processes, developed agricultural practices that suppressed pests. For example, crop rotation and application of manure were

---

Experimental Demonstration—continued

but increased greatly in treated plots because the fungicide killed the nematodes' biological-control organisms (Stirling, 1991).

Soil-Borne Pathogenic Fungi

Certain soils are suppressive to specific soil-borne fungi, such as *Gaeumannomyces graminis* var. *tritici*, which causes take-all of wheat, or *Fusarium oxysporum*, which causes Fusarium wilt of many crops. Although propagules of a pathogenic fungus may be present in these suppressive soils at densities adequate to cause disease, disease is absent or symptoms are very mild. For example, if a suppressive soil is inoculated with propagules of *G. graminis* var. *tritici*, symptoms of take-all of wheat typically are very mild. In contrast, if the suppressive soil is fumigated or pasteurized prior to inoculation with the pathogen, symptoms of take-all are severe (Cook and Weller, 1987). Similarly, the suppressive effect of Fusarium-suppressive soils is destroyed by biocidal treatments and can be restored by mixing a small quantity of suppressive soil into a pasteurized soil (Alabouvette, 1993). Disease suppression in both cases is attributed to interactions between the pathogen and the saprophytic microflora present in suppressive soils.

Weed Composition Shifting

Weed-shifting is a phenomenon in which previously innocuous plants and minor weeds emerge as dominant weeds following the removal of competition from erstwhile major weeds. Numerous instances are reported of differential susceptibility of weed species to herbicides. Continued use of a particular herbicide often causes a shift within a weed community from susceptible to more tolerant species. For example, a species shift that favors grasses is readily observed from applications of 2,4-D for broad-leaf weed control in cereals. Other examples include the increased occurrence of pigweed (*Amaranthus* spp.) from napropamide, mustards (*Brassica* spp.) from benefin, and common groundsel (*Senecio vulgaris*) from diuron and terbacil (Radosevich and Holt, 1984). The phenomenon of weed shifting demonstrates that certain weed populations are suppressed by more dominant populations, in the absence of herbicides.

---

used in China approximately 3,000 years ago. In the first century B.C., fallowing fields with poor crop yields was a recommended practice (Cook and Baker, 1983). Similarly, fallowing, rotating cereal crops with legumes, composting, manuring, liming, and optimizing schedules of irrigation and drainage were implemented throughout the Roman empire (50 B.C.–476 A.D.) (Cook and Baker, 1983).

Throughout the history of agriculture, the benefits achieved through advocated cultural practices were recognized, but the importance of natural processes in realizing many of these benefits only began to be appreciated when science discovered microorganisms in the mid 1800s. The phenomenon of antagonism (a microorganism's ability to sustain its life by parasitizing another organism) was recognized in the late 1880s, but its occurrence in agricultural soils and its role in suppression of plant pathogens and other pests was not fully recognized until early in this century. It is now firmly established that many practices implemented by early agriculturalists enhance the diversity and activities of soil microorganisms, including biological-control organisms. Ironically, the importance of natural processes performed by indigenous biological-control organisms are demonstrated most clearly when they are destroyed by chemical or physical means. If indigenous biological-control organisms are destroyed by sterilization or fumigation of soil, for example, an unaffected indigenous or an introduced pathogen can cause far more damage than would be possible in untreated soil. Many of the cultural practices developed throughout the evolution of agriculture are still used today.

Crop rotation (successive planting of different crops in the same field) is a proven cultural strategy to suppress weed, arthropod, and pathogen pests (National Research Council, 1989b). Optimizing the benefits of crop rotations requires knowledge of the effects of diversified plantings on a cropping system. Rotating a primary crop with another cash crop can limit growth of a pest that feeds on the primary crop (National Research Council, 1989b). Root food sources of soil-borne arthropods as well as inoculum densities of plant pathogens are severely reduced by crop rotation. For instance, populations of plant-parasitic nematodes—a persistent problem in continuous soybean cropping systems—are nearly eliminated by switching to corn in alternate growing seasons. Crop rotation alters a soil environment, which can reduce the devastating effects of many soil-borne pests.

Other cropping practices used to encourage growth of beneficial organisms and suppress pest populations include use of green manures (a legume crop plowed under to increase soil fertility), cover crops (crops grown for ground cover), and intercropping (interspersing crops). An example of intercropping is the use of trap crops grown in rows beside a primary crop to both provide habitat for beneficial arthropods and divert pest arthropods from preying on the primary crop. Knowing the effects of cropping on ecosystem components can lead growers to a diversity of plant resources useful in crop-production systems.

A combination of cultural practices is synergistic and can be more effective than a single tactic alone against an agricultural pest. For example, against weeds, a grower may cultivate a field to prevent germination of weed seeds, then plant the crop to ensure that the crop has a competitive advantage in access to limited water, light, and nutrient sources. Manipulation of planting and harvesting schedules also can have a negative impact on pest populations (National Research Council, 1989b; Ferris, 1992; Schroth et al., 1992); however, optimizing these schedules requires prior knowledge of crop-arthropod life cycles and host-free stages of the pest (Edwards and Ford, 1992). With knowledge of ecological interactions, a grower can integrate several cultural practices to manage a variety of agricultural pests.

PLANT BREEDING

Improving crops by selective breeding—choosing to plant the seed of the largest fruit or the most easily harvested plants, for example—started thousands of years ago (National Research Council, 1984). In 1866 Mendel published findings on controlled pollination techniques which demonstrated that plant characteristics are inherited in a predictable manner. Since 1900 knowledge of genetics has been used to continuously improve crop varieties. The release of new varieties of hybrid corn in 1930 and a rust-resistant wheat in 1938 led to major gains in crop yields per acre.

Breeding to increase plant vigor and resistance can be achieved by classical and molecular approaches; both approaches modify plant DNA and transfer desirable agronomic traits. A common classical technique, hybridization, is used to transfer genes from one plant to another compatible plant of the same genus or species. Crop improvement through hybridization, however, can be limited by a low rate of reproduction from progeny with the desired trait; and with only one growing season per year for many agricultural crops, many years are needed to develop a new cultivar. Nonetheless, classical breeding techniques continue to be the primary way to develop and improve plants to meet agricultural needs (National Research Council, 1989a,b).

Breeding plants to increase their ability to resist attack by pests continues to be an essential component of pest management. Breeding for a uniform trait, however, is not without risk; genetic uniformity can have severe negative risks. An example occurred in 1970 when more than 90 percent of the U.S. corn crop was planted with corn developed from one source of mass-produced hybrids carrying a trait called the Texas male-sterile (Tms) cytoplasm. Corn derived from the Tms hybrids facilitated corn breeding because it was male sterile and therefore did not require tedious and expensive detasselling. However, Tms hybrids were extremely susceptible to race T of *Bipolaris maydis*, the causal agent of southern corn leaf blight. Race T had always been present but was relatively unimportant until the Tms hybrid corn was planted, almost exclusively. This large-scale planting of a highly susceptible corn hybrid led to a severe outbreak of southern corn leaf blight (Poehlman, 1979). The following year plant breeders returned to improved, inbred lines of corn containing normal cytoplasm, and the disease abated.

A molecular approach to breeding plants with improved agronomic traits involves transgenic techniques using molecular technology. Such techniques involve transferring foreign DNA via a biological agent or by physically inserting the foreign gene into the plant cell. Transgenic techniques can be more precise than hybridization because a specific gene is transferred, and the time required to

Take-All Disease and Its Decline

Take-all is a major disease of wheat worldwide. The soil-borne fungus *Gaeumannomyces graminis* var. *tritici* infects wheat roots, making the infected roots less efficient in taking up water and nutrients from the soil and in transporting these to the aerial portions of the plant. As a result, infected plants are stunted and often display premature senescence. Severe infections can kill plants in the seedling stage, but the disease typically develops more slowly and plants die after heading but before mature grain can be harvested. Crop rotation is an effective means of managing the disease because the pathogen does not persist in soil for extended periods in the absence of wheat plants. However, in many regions where wheat is grown, because of economic pressures, growers plant consecutive crops of wheat without rotation.

More than 50 years ago, scientists observed that although the severity of take-all increased with each consecutive crop of wheat for approximately 4 years, continued monoculture of wheat beyond 4 years often resulted in decreased disease severity. Scientists attributed this phenomenon, called "take-all decline," to the proliferation of antagonistic bacteria on wheat roots and in infested root debris. In fields displaying take-all decline, bacterial antagonists, primarily fluorescent pseudomonades, suppress the growth of *G. graminis* var. *tritici* on root lesions and in crop debris. Certain strains of fluorescent pseudomonades isolated from soil in fields where take-all has been naturally suppressed produce antifungal compounds toxic to *G. graminis* var *tritici*. In field experiments, when inoculated onto seeds, fluorescent pseudomonads suppress take-all and increase yield of wheat.

The take-all decline system illustrates a form of biological control that is achieved through the cultural practice of wheat monoculture, which causes a shift in the composition of microorganisms present on the roots and in crop debris. Although crop rotation is an invaluable agronomic practice for the management of soil-borne pathogens including *G. graminis* var. *tritici*, monoculture of wheat favors the proliferation of bacteria suppressive to the take-all fungus. Take-all decline is a well-studied example of what may be a common phenomenon: that much of the value of cultural practices may be achieved through their effects on the microorganisms that form an essential component of the agricultural ecosystem.

---

develop a viable, commercial cultivar can be decreased (National Research Council, 1989a).

Genes from animal, microbial, or plant sources can also be inserted into plants to produce a variety with novel resistance characteristics. The first example of genetic engineering for disease resistance involved a gene encoding for the coat protein of the tobacco mosaic virus (TMV) that was introduced into tobacco plants through an agrobacterium vector (Powell-Abel et al., 1986). These transgenic plants were resistant to TMV as well as other related viruses. Modern tools have enabled scientists to enlarge the pool of genetic resources to make needed crop improvements.

Expression of coat protein genes have become increasingly important in developing new varieties of resistant crops; since 1986 there have been more than 75 reports of transgenic plants with virus resistance as a result of the expression of coat protein genes in crops as varied as potato, tomato, squash, rice, corn, sugar beets, and lettuce (Fitchen and Beachy, 1993). In 1995, the first commercial variety of such virus-resistant plants was introduced—crooked-neck squash resistant to cucumber mosaic virus and zucchini yellow mosaic virus. More recently, genes for resistance to a number of different diseases were isolated from tomato, tobacco, Arabidopsis, flax, and rice plants, to name a few. Whitham et al. (1994) isolated a gene for resistance to TMV; Johal and Briggs (1992) isolated a gene for resistance to corn blight. In both cases, resistance was achieved when the resistance gene was reintroduced into a previously sensitive plant variety.

SYNTHETIC ORGANIC PESTICIDES

Between World Wars I and II, large-scale production practices coupled with major developments in synthetic chemistry revolutionized the pesticide industry. Tetraethylpyrophosphate (TEPP), the first organophosphate insecticide, was synthesized in 1938; DDT followed in 1939. An analog of the plant hormone indole acetic acid was synthesized in 1939 as 2,4-D and introduced in 1942 as the first synthetic selective herbicide. The first dithiocarbamate fungicide, zineb, was introduced in 1943; and in 1945, chlordane, the first of the persistent, chlorinated insecticides, and propham, the first carbamate herbicide, were introduced. In 1947, toxaphene, which became the most heavily used insecticide in the United States, was introduced, followed by aldrin and dieldrin in 1948 and malathion in 1950.

In the 1940s and 1950s, soil fumigants, including methyl bromide, ethylene dibromide, dichloropropenes, and dibromochloropropane (DBCP), were found to effectively suppress soil-borne, parasitic nematodes and fungi. New insecticides, herbicides, fungicides, and nematicides were introduced almost annually through the 1950s and 1960s. As remarkable as the pace of discovery and development was the rate of adoption of the new pesticides by growers (Osteen and Szmedra, 1989), who found that the new chemicals (1) were highly effective and predictable at reducing pest populations, (2) produced rapid and easily observed mortality of the pest, (3) were flexible enough to meet diverse agronomic and ecological conditions, and (4) were inexpensive treatments compared to the crop damage that would be otherwise sustained (Metcalf, 1982; National Research Council, 1975). Synthetic pesticides quickly became the favored means of crop protection and dramatically eclipsed other approaches to pest management.

The ready acceptance of pesticides, coupled with comparable increases in the use of commercial fertilizers and mechanization, revolutionized agricultural production. Yields have increased steadily and dramatically since the 1940s, and specialized production of one or a few crops has become the norm for growers.

Insecticides

By the 1950s, insecticides were used broadly on high-value crops especially susceptible to arthropod attack: cotton (66 percent treated, by area), fruits and nuts (81 percent), potatoes (80 percent), vegetables (74 percent), and tobacco (58 percent). During the 1960s and 1970s both herbicides and insecticides were used on corn and tobacco, and use was expanded to include other major field crops such as cotton, soybean, sorghum, rice, peanuts, wheat and other small grains, hay, and pasture (Osteen and Szmedra, 1989).

From 1964 to 1982 the total volume of insecticide use on cotton declined (Osteen and Szmedra, 1989) after the introduction of pyrethroids, a pesticide used at much lower rates but applied more frequently (Zalom et al., 1992). During this same period, the volume of insecticide applied to corn and soybean nearly doubled. In 1982 the amount of insecticide used on soybean and corn exceeded that used on cotton for the first time in the United States (Osteen and Szmedra, 1989).

Herbicides

During the 1960s and 1970s, increased herbicide applications on corn and soybean accounted for the largest pesticide use increases (Lin et al., 1995). By the 1980s herbicides accounted for 75 percent of total volume of pesticides applied to agricultural crops. The total amount of herbicides used decreased in the 1990s in part because newly introduced herbicides were used at lower application rates. Rice crops, though their total acreage is small, receive the most intensive use of herbicides, 2.5 kg/hectare (5.6 lb/acre). Herbicides account for more than 90 percent of pesticide applications in corn and soybean production today (Lin et al., 1995).

Declining Use

Since peaking in 1982, pesticide use declined from 270 million kg (600 million lb) in 1982 to 260 million kg (570 million lb) in 1992. Currently, corn and soybean production, occupying the largest percentage of crop acreage in the United States, dominate pesticide use. On a per-acre basis, however, pesticides are used more intensively on many fruit and vegetable crops, including potatoes, than on corn; for example, fungicides applied to citrus and apple trees (3.6 million kg or 8 million lb) account for 23 percent of all fungicides used. Cotton accounts for about 10 percent of all pesticides used. Pesticide applications on wheat, a major crop with the second largest acreage in the United States, account for only 3.5 percent of the total (Lin et al., 1995).

INTEGRATED PEST MANAGEMENT

Integrated pest management (IPM) was intended to put pesticide use on a more sound ecological footing. Stern and colleagues (1959) introduced the term *integrated control* and defined it as "applied pest control which combines and integrates biological and chemical control. Chemical control is used as necessary and in a manner which is least disruptive to biological control" (p. 86). Pest management was to be based on the naturally occurring regulatory processes in the agroecosystem that help prevent pest outbreaks.

Stern and colleagues (1959) suggested that pests should be reduced (as opposed to completely exterminated) only when their populations reach levels at which economic injury to the crop is expected, thus also introducing the concepts of economic-injury level and economic threshold to guide decisions about when and whether pest populations should be reduced. *Economic-injury level* is determined by measuring the effect on the crop plant and the damage it can tolerate. *Economic threshold*, set below the economic-injury level, serves to signal the need for action to keep the pest population from reaching the point at which economic injury would occur.

The founding principles of IPM are that natural processes can be manipulated to increase their effectiveness, and chemical controls should be used only when and where natural processes of control fail to keep pests below economic-injury levels. These principles were refined, expanded, and incorporated into many of the concepts of pest management advocated by others (Doutt and Smith, 1971; National Research Council, 1969; Rabb and Guthrie, 1972; Smith, 1969; Smith and van den Bosch, 1967; U.N. Food and Agriculture Organization, 1967). Similar concepts and definitions of pest management were used as the basis for accelerating adoption of national agricultural initiatives during the administrations of presidents Nixon (Council on Environmental Quality, 1972) and Carter (Bottrell, 1979).

IPM in its original sense of integrated control or ecologically based management, however, has not been implemented on a wide scale (Cate and Hinkle, 1993; Flint and van den Bosch, 1981; Hoy and Herzog, 1985; Kogan, 1986). Critics argue that the most widely used IPM strategies stress improved pesticide usage based on monitoring pest populations and setting economic thresholds (Fitzner, 1993). Many scientists have noted that IPM strategies normally depend on pesticides as the primary management tool and have highlighted the need to develop systems that depend primarily on biological control organisms, resistant plants, cultural controls, and other ecologically based tools (Cate and Hinkle, 1993; Edwards, 1991; Ferro, 1993; Flint and van den Bosch, 1981; Frisbie et al., 1992; Frisbie and Smith, 1989; Hoy and Herzog, 1985; Kogan, 1986; Pedigo and

Higley, 1992; Tette and Jacobsen, 1992; Zalom and Fry, 1992). Indeed, some scientists have proposed that the term *biologically intensive IPM* be used to distinguish IPM strategies that rely on biologically based tools from those that depend primarily on conventional broad-spectrum pesticides (Edwards, 1991; Ferro, 1993; Frisbie and Smith, 1989; Pedigo and Higley, 1992; Prokopy, 1993; Zalom and Fry, 1992). The historical emphasis on arthropod control, and the subsequent lower priority of pathogen and weed management, does create some confusion for IPM practitioners seeking environmentally sound and economically feasible answers to pest management problems.

Naturally occurring compounds, biological-control organisms, and resistant plants have been developed and used for most of the history of pest management. With the advent of synthetic chemical pesticides, emphasis in research and practice shifted away from biologically based strategies. However, contemporary advances in scientific knowledge coupled with the long experience of the past provides a solid foundation for renewed effort in identifying appropriate ecological approaches to pest management.

OBSTACLES TO CONTINUED USE OF
BROAD-SPECTRUM PESTICIDES

Pesticides were readily accepted by growers because, at least initially, they were quite successful at suppressing pests. However, problems began shortly after pesticide use became widespread. Arthropod resistance to DDT was first observed in Sweden in 1946, only 7 years after DDT was introduced. By 1948, 14 species of arthropods were reported to be resistant to DDT, cyclodienes, organophosphates, carbamates, or pyrethroid insecticides; that number exceeded 500 by 1990 (Gould, 1991). But the problem is not simply that some pests *develop* resistance; some were never controlled by pesticides. For some soil-borne pathogens, nematodes, arthropods, and aquatic weeds, there are no acceptable conventional chemical pesticides.

Widespread use of pesticides has also raised concerns about the health effects of pesticide residues in foods humans and livestock animals eat. In 1954, the Food, Drug, and Cosmetic Act was amended to set maximum tolerance levels of pesticide residues in raw agricultural commodities because of concern about health effects and dietary exposure.

Problems and Limitations of Pesticides

The early success of broad-spectrum, synthetic pesticides raised hopes that pest problems that had plagued agriculture from its inception had finally been solved. The inability to sustain the initial effects achieved by conventional pesticides, however, is an important reason to be concerned about continued dependence on ineffective pesticides that may also adversely impact nontarget species in the environment. The development of resistance to pesticides, the increasing cost of discovering and developing pesticides, and the fact that some pest problems have been caused or exacerbated by the use of pesticides have all contributed to arguments for decreased reliance on these chemicals to provide crop protection.

Development of Resistance

Major arthropod agricultural pests have developed resistance to more than one class of insecticides (Georghiou, 1986; Gould, 1991) (Figure 1-1). Growers normally respond to the first signs of resistance by increasing their rate of application and then, if that does not work, switching to another pesticide that may also become ineffective. The problem is further exacerbated by the ability of a given population of pests to develop several different mechanisms of resistance with each mechanism responding to a different pesticide (Georghiou, 1986).

The increased cost of continued pest management resulting from pesticide resistance can be substantial. Georghiou (1986) cited examples including a more

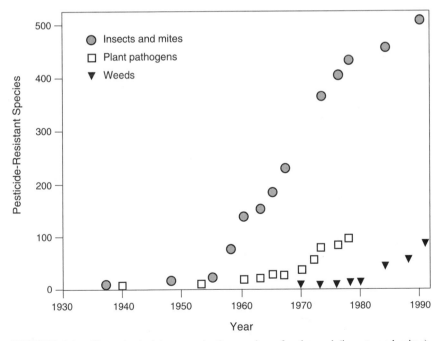

FIGURE 1-1 Chronological increase in the number of arthropod (insects and mites), plant pathogen, and weed species resistant to synthetic chemical pesticides since their development in the late 1930s. SOURCE: Adapted from Gould, F. 1991. The evolutionary potential of crop pests. Am. Sci. 79:496–507.

than 5-fold increase in the costs of malaria control when DDT was replaced by malathion and a 15- to 20-fold increase when malathion was replaced by propoxur, fenitrothion, or deltamethrin (Metcalf, 1983). In the early 1990s it was estimated that development of resistance added $400 million to the cost of pest management in the United States (Pimentel et al., 1992).

Since systemic fungicides were first introduced for agronomic use in the 1960s, their use to control fungal diseases of plants has been compromised by the development of resistance to these compounds (Dekker, 1993). For that reason, antibiotics and copper compounds used for management of plant diseases are no longer effective against many bacterial plant pathogens (Jones, 1982). In many cases, antibiotic resistance is conferred by plasmids transferred among bacterial species, thereby enhancing the spread of antibiotic or copper resistance among bacterial plant pathogens (Cooksey, 1990).

As regards herbicides, repeated use has caused plants, when reproducing, to select for herbicide-resistant biotypes with alterations in metabolic enzymes targeted by herbicides. Holt and LeBaron (1990) listed 55 weed species, including 40 dicots and 15 grasses, that had biotypes resistant to triazine herbicides. They reported that one or more resistant species had developed in 31 states of the United States and in 4 provinces of Canada. Resistance to several classes of herbicides, such as triazine, trifluralin, paraquat, dichloropmethyl, substituted urea, and sulfonylurea, is rapidly becoming a worldwide problem. Inasmuch as some weed biotypes have developed resistance to several classes of herbicides, herbicide resistance appears to have reached the level where concerted efforts are needed for effective management strategies (Holt and LeBaron, 1990).

The appearance of new pesticide-induced pests and the development of resistance puts growers on a pesticide treadmill—applying pesticides at ever increasing rates as resistance develops or as secondary pests emerge and eventually switching to a new chemical, often at higher cost. Continued reliance on pesticides makes it impossible to step off the treadmill. Experience suggests that resistance should be anticipated as a possible outcome of pest control; thus, management approaches must be developed in tandem with pesticide deployment to minimize resistance problems.

Escalating Costs and Fewer Discoveries

The cost of developing new agricultural chemicals has increased dramatically since 1956. According to the American Crop Protection Association, nearly 12 percent of proceeds from agricultural sales in the United States are spent on research and development, totaling more than $650 million. Despite this investment, the probability that a successful commercial product will be discovered has declined. In fact, since 1970 there has been a steady decline in the number of new pesticides introduced (Ollinger and Fernandez-Cornejo, 1995). Consequently, synthetic chemical pesticide companies are consolidating to maintain profitability as discovery declines and development costs increase (Ollinger and Fernandez-Cornejo, 1995).

Pesticide-Induced Pest Problems

Broad-spectrum insecticides that kill nontarget arthropods can exacerbate or even create new pest problems by eliminating biological control organisms that previously held the pests in check. For example, DeBach and Rosen (1991) described the elevation of the citrus red mite from a minor or nonpest status to the most important citrus pest in California following the introduction of chlorinated hydrocarbon and organophosphate pesticides. Similar increases in mite problems after the introduction of pesticides have been reported worldwide (Gerson and Cohen, 1989; Huffaker et al., 1969, 1970; McMurty et al., 1970). Cottony-cushion scale emerged as a major pest in California's central valley after DDT caused large reductions in the populations of a biological control organism (DeBach and Rosen, 1991); brown soft scale became a major citrus pest in the Lower Rio Grande Valley in Texas in 1959 when parathion applied to adjacent cotton fields drifted into citrus groves and killed biological control organisms (Dean et al., 1983); and, outbreaks of citrus red mite, purple scale, and woolly whitefly followed large applications of carbaryl and chlordane in an attempt to eradicate the Japanese beetle (DeBach and Rose, 1977).

Diseases induced by, or made more severe after, the use of healing agents are termed "iatrogenic" diseases; examples of iatrogenic plant diseases can be cited for all major groups of crop protection chemicals (Griffiths, 1993). Botrytis rot of cyclamen (caused by *Botrytis cinerea*) was initially well controlled by the fungicide benomyl, but it became much more severe when benomyl-resistant strains appeared. Prior to the use of benomyl, populations of *B. cinerea* were suppressed by antagonistic strains of *Penicillium brevicompactum*. Benomyl eliminated the *P. brevicompactum*, which was sensitive to the fungicide, so that when the benomyl-insensitive strain of *B. cinerea* appeared, it was more damaging to the host than it had been prior to the discovery of benomyl. Eventually, benomyl-insensitive strains of *P. brevicompactum* appeared, effectively returning the original balance of biological suppression (Griffiths, 1993).

Unjudicious use of broad-spectrum chemical pesticides for pest management has encouraged resistance in agricultural pests, increased secondary pest problems, increased the probability of dietary and occupational exposure to harmful chemicals, and produced adverse effects on soil and water resources and nontarget species.

Problems that Defy Conventional Chemical Solutions

For many pest problems, no conventional pesticide offers a feasible solution. Examples include problems caused by exotic pests (for which there are no indig-

Management of Arthropod Pests in Cotton Production

The history of pesticide-induced pest problems in cotton production serves as one of the most compelling examples of the limitations of broad-spectrum chemical pesticides to provide long-term pest control. The development of resistance to pesticides by primary cotton pests such as the boll weevil combined with simultaneous pesticide-induced outbreaks of secondary pests led to a dramatic reduction in the early 1980s in acreage planted to cotton in the southeastern states and threatened to curtail cotton production in the Rio Grande Valley of Texas, the Imperial and Central Valleys of California, and in various Central American countries (Carlson et al., 1989; DeBach and Rosen, 1991).

Boll Weevils vs. The Chemicals

The history of attempts to manage the boll weevil (*Anthonomous grandis*) in the Cotton Belt is a classic example of the pesticide treadmill. The standard treatment from about 1910 through 1949 was multiple applications of calcium arsenate dust. This method was quickly replaced in 1950 when chlorinated hydrocarbon compounds such as DDT, aldrin, and dieldrin were introduced along with broad-spectrum pesticides such as toxaphene. These materials, applied either alone or in combination, were successful in managing the boll weevil—and its natural predators—until 1954 when the boll weevil's development of resistance to DDT was first reported (Johnston, 1961; Smith et al., 1964).

Between 1956 and 1958, organophosphates, particularly methyl parathion, and several others such as malathion, methomyl, azinphos-methyl, carbaryl, endrin, and heptachlor, were introduced to replace DDT and other materials that had become ineffective in the battle against the boll weevil. Widespread use of these toxic, broad-spectrum insecticides eliminated the natural organisms that controlled bollworm (*Heliothis zea*) and tobacco budworm (*H. virescens*). Populations of these formerly minor pest species increased, which contributed to additional serious economic losses. DDT was then reintroduced to control bollworm, but by as early as 1962 this pest became resistant to DDT and to other chlorinated hydrocarbon and carbamate insecticides. Growers then switched back to methyl parathion; again, resistance soon developed, and by 1968 it was not uncommon for growers to apply 15 to 18 treatments per season without achieving satisfactory insect control. Rainwater (1962) estimated that one-third of all insecticides used for agricultural purposes were being applied for control of cotton pests, primarily the boll weevil. Bollworms and budworms, which had reached pest status through the use of insecticides applied to control the boll weevil, now rivaled the boll weevil as the most serious pests in cotton production.

Fortunately, pioneering work by Brazzel and colleagues (1961) showed that a series of four malathion pesticide treatments applied in the fall would reduce by 90 percent the population of boll weevils emerging in the spring. The effectiveness of this new management procedure was demonstrated in the Texas High Plains-

Management of Arthropod Pests—continued

suppression program of 1964-1965 when 1.1 million acres of cotton was treated four times in the fall with malathion applied as an ultra-low volume spray. Knipling (1968, 1971) later reported that seven treatments applied in the fall would increase suppression of spring boll weevil populations to 98 percent, especially when used in combination with cultural practices and pheromone-trapping technology. This work led to the initiation of a successful interstate coordinated effort, the Pilot Boll Weevil Eradication Experiment, in southern Mississippi and adjoining areas of Louisiana and Alabama from 1971 to 1973. The favorable results of the Mississippi Pilot Experiment indicated that elimination of the boll weevil from the continental United States was technologically and operationally feasible, and the first full-scale boll weevil eradication program was initiated in North and South Carolina in 1983. The program was expanded to include western Arizona, southern California, and northern Mexico in 1985-1986; Georgia, Florida, and parts of Alabama in 1987; and the remainder of Arizona in 1988 (Brazzel, 1989). Since initiation of the eradication program in 1983, the boll weevil has been virtually eliminated from large sections of the southeast.

The Success of a Combined-Treatment Strategy

The elimination of this primary pest together with development of an IPM program for other cotton pests has served as a major stimulus for the restoration of economic cotton production in the Cotton Belt. The IPM program is based on biology and behavior, monitoring, cultural practices, and prudent pesticide application. It has provided several significant benefits, including savings in pesticide and pest-control costs, increased yields, substantial increases in acreage planted in cotton, and increased land value (Carlson et al., 1989; Cate, 1988; Roach et al., 1990). In Georgia, for example, pesticide treatments have been reduced from an average of 17 applications/acre/year to 4 applications/acre/year (Suber and Todd, 1980), while overall pest-control costs and losses from damage have been reduced by 50 percent. The number of treatments for control of bollworm and budworm, mites, aphids, and plant bugs has declined as beneficial species that were previously suppressed by insecticides reemerge as effective biological-control agents. In general, per-acre yields have nearly doubled since initiation of the eradication program. Cotton acreage in Georgia has increased from an historic low of 115,000 acres in 1983 to 875,000 acres in 1994, with planting estimates of 1.25 million acres for 1995. The 1994 yield of 1,480,000 bales was the highest in Georgia since 1935, when cotton was produced on more than 2.5 million acres. Total value of the crop in 1983 was $44 million vs. $497 million in 1994. Other discussions of pesticide savings, enhanced yield, and increases in acreage planted following the implementation of durable pest management and boll weevil eradication in cotton have been documented in studies by Cate (1988), Carlson et al. (1989), and Roach et al. (1990).

enous biological controls), soil-borne arthropods, nematodes, and pathogens, plant viruses, rangeland weeds, row-crop weeds, and aquatic weeds.

Invasion of Exotic Pests

The U.S. Office of Technology Assessment (1993) estimated that more than 4,000 plants, arthropods, and plant pathogens of foreign origin have established free-living populations in the United States. Some once-exotic species clearly are beneficial; soybean and wheat are now primary crops throughout much of the United States and staples in the U.S. diet. Most exotic pests arrived in the United States through human activity and were released, unintentionally or deliberately, before they were recognized as pests. Some well-known exotic pests include

- arthropods such as Mediterranean fruit fly, boll weevil, gypsy moth, pink bollworm, Japanese beetle, European corn borer, sweet potato whitefly, and Russian wheat aphid;
- pathogens such as white pine blister rust, Dutch elm disease, chestnut blight fungus, and potato blight fungus; and
- weeds such as hydrilla, water hyacinth, purple loosestrife, leafy spurge, and kudzu.

Losses to agricultural production caused by exotic plant pests alone equal approximately $28.8 billion per year, and expenditures for their prevention and control equal approximately $3.2 billion per year (Schwalbe, 1993).

The difficulty in identifying an exotic pest species in a timely manner can delay deployment of natural predators. Classical biological control efforts to solve exotic pest problems have not had a high success rate, suggesting an inadequate ecological basis for this approach. A better understanding of exotic pests and their natural predators and parasitoids will increase the potential of biological control as a viable approach for management of these pests.

Soil-Borne Pathogens, Nematodes, and Arthropods

Soil-borne plant diseases are caused by pathogenic fungi, viruses, viroids, mycoplasmas, bacteria, or nematodes that live in or on the surface of soil. No practical chemical controls are available for many soil-borne diseases, including *Pythium*, *Rhizoctonia*, and *Phytophthora* root rots; *Fusarium* wilt; root-knot, cyst, and other nematodes; or arthropods such as the Japanese beetle and root weevils. Losses from root diseases cause wheat farmers an estimated $1.5 billion annually (U.S. Department of Agriculture, Agriculture Research Service, 1995). There are two primary reasons why these problems cannot be controlled by current pest management strategies.

1. It is difficult to achieve uniform coverage of the root system with non-

Soybean Cyst Nematode—An Unmet Need

The soybean cyst nematode (*Heterodera glycines*) (SCN) has been a persistent problem in soybean production regions throughout the South and Midwest and caused more than $250 million in yield reduction in 1991 alone (Meyer and Huettel, 1993); over the last 20 years, it has been the most important soil-borne pest in soybean production in the southern United States (Wrather et al., 1995). Management tools have included nematode-resistant cultivars, nematicides, and cultural controls (Riggs and Wrather, 1992). These tools have been useful but have significant problems or have been inadequate by themselves; new biological control tools will help combat this pest.

Although nematode-resistant cultivars have been useful in combating SCN, resistant cultivars often produce lower yields than do susceptible cultivars. More important, the evolution of "resistance-breaking" nematode races reduces the value of resistance cultivars. Initially, the resistant cultivar is quite effective, as it prevents reproduction of the dominant race of SCN in the particular growing region. But SCN consists of many races, and some developed the ability to reproduce on the resistant cultivars. These resistance-breaking races may be present at low levels initially, but they increase to damaging levels when the resistant cultivar is planted repeatedly. Alternating cultivars with different sources and levels of resistance can reduce the chance that resistance-breaking races will increase to damaging levels (Ferris, 1992).

Nematicides, including soil fumigants, effectively control nematodes but have serious limitations. First, they are increasingly unavailable to growers; several effective chemicals, including dibromochloropropane (DBCP) and ethylene dibromide (EDB), have been cancelled because of concerns regarding contamination of groundwater as well as other environmental and toxicological concerns. Second, nematicides are costly. Although application of nematicides often is profitable on high-value crops such as tomato, pepper, eggplant, and strawberry, application often is not cost effective in lower-value crops. In the case of soybeans, nematicides were applied extensively in the 1970s and 1980s when the value of the crop was high, but nematicides have seldom been applied in recent years because the value of the crop is lower.

Cultural controls include crop rotation, which is effective because SCN does not have a broad host range and does not persist in soil for extended periods of time. Rotation to nonhosts, however, costs the grower because the nonhost crops (corn, wheat, other grasses, etc.) are lower in value than soybean. Management of weed hosts and management of resistant cultivars, to avoid the race shifts, also are important.

In addition to SCN, other plant-parasitic nematodes (including lesion, ring, sting, stunt, and especially root-knot nematodes) damage a wide variety of crops and are difficult to control. Fortunately, these nematodes have a number of natural enemies, and biological controls are being investigated in several laboratories in the United States and elsewhere. The fungus *Verticillium lecanii*, in combination with a sex pheromone, reduced SCN in recent greenhouse and field tests (Meyer and Huettel, 1993). Another fungus, as yet unnamed, occurs naturally in certain soybean production regions in Arkansas and maintains SCN at low levels (Kim and Riggs, 1994). Researchers are studying these and other beneficial organisms to determine how they can be integrated with resistance and cultural controls to manage SCN and other nematode pests.

systemic fungicides. It is especially difficult to protect root tips, which are the site of infection for many soil-borne pathogens.

2. Many fumigants such as dibromochloropropane and ethylene dibromide, two of the most effective nematicides, have been removed from the market; and the cost of using those fumigants that remain on the market often is too great to justify their use on field crops such as cereal and oil crops (Cook et al., 1995).

Plant Viruses

Plant viruses annually cause crop-yield reductions valued at more than $100 million. In most cases the reductions represent 1 percent to 10 percent of the potential crop yield, a figure generally accepted in cases where disease resistance is not available. In selected crops the impact can be severe, causing losses of up to 80 percent. As with human medicine, there are no chemicals capable of controlling plant viruses—chemical controls must be directed at the arthropod, fungal, or nematode vectors of plant viruses. Control of plant viruses is effected by

- management of areas surrounding production areas,
- control of the insects and other vectors of the disease agent, and
- use of disease resistant varieties.

Agriculturalists have long known that the incidence and severity of virus diseases can be reduced by

- eliminating weeds that harbor viruses and serve as alternate hosts from which insect vectors spread the virus;
- using insecticides to control the vectors, for example, aphids and white-flies;
- crop rotation and crop sanitation practices to reduce the sources of inoculum; and
- shifting planting dates to avoid infestations of insect vectors.

Even with recommended agricultural practices, growers face the possibility of substantial losses caused by disease in years in which climactic changes augment heavy infestations of arthropod vectors. For example, outbreaks of wheat streak mosaic virus can occur in the Midwest in the spring following a warm, wet fall season—conditions optimal for the mite vector.

The situation is not as bleak with all crops, however, largely because of the success of plant breeders in identifying and deploying natural genes for resistance. Genes for resistance to tomato mosaic virus; potato viruses X, Y, and M; cucumber mosaic virus in pepper and in some varieties of cucumber, tomato, and cantaloupe are all examples of successful deployment of genetic material. Some but not all virus resistance genes are single genetic loci, easily transferred between varieties by cross-breeding. Other resistance genes are multigenic and not easily transferred.

Plant breeders have also identified genes that confer appetite suppression to the insect vector: reduced feeding leads to reduced transmission and disease. Coupling appetite suppressant genes with genes that confer resistance to the virus can produce excellent control of viral diseases. Unfortunately, it has not been possible to identify and successfully deploy resistance genes to control the more than 50 significant diseases caused by viruses in agriculture in the United States, Canada, and Mexico.

Rangeland Weeds

Rangelands comprise about 312 million hectares (770 million acres) in the United States (National Research Council, 1994). These diverse ecosystems serve many purposes including a major agricultural use as grazing lands for livestock. Annual weeds often dominate because of their faster growth rate in comparison to desirable perennial grasses or, as in the case of broad-leaved weeds, are unpalatable and easily spread on improperly grazed lands (National Research Council, 1975). The vastness and lack of accessibility of these remote land parcels make them particularly vulnerable to introduction of exotic weed species and shifts to nonforage native species (National Research Council, 1975; U.S. Office of Technology Assessment, 1993). Since herbicides are costly and such large land areas are involved, solving these weed problems becomes an enormous task (National Research Council, 1975; Turner, 1992; U.S. Office of Technology Assessment, 1993). Nevertheless, opportunities may exist to apply a combination of innovative biological, chemical, physical, and cultural techniques to manage rangeland weeds.

Row-Crop Pests

Pest management in row crop production systems represents a major challenge. The value of row crops—annual crops grown in rows (vegetables and small fruit)—is estimated to be more than $7.5 billion annually (Zalom and Fry, 1992). Generally, many row-crop growers depend on methyl bromide, a highly effective fumigant used against soil-borne pathogens and arthropods as well as some perennial and annual weeds (U.S. Department of Agriculture, 1993), to provide the high-quality, good-tasting, unblemished produce consumers demand. In addition, some herbicide sprays can damage nearby seedlings as well as weeds. Because acreage allotted to row crops is usually less than that allotted to major grain crops, pesticide manufacturers have little incentive to invest heavily in research and development of new products for row crops. Thus there is an urgent need for new pest management strategies for these horticultural systems.

Aquatic Weeds

In North America more than 170 species of aquatic plants are classified as weeds; 40 to 50 species are considered to be of major importance (Andres and Bennett, 1975). The managing aquatic weeds is complicated by the fact that unrelated aquatic plant species, including some native species that are beneficial, may coexist with aquatic weeds; and not all weed species may require control in every situation. Moreover, water bodies are typically subject to multiple uses and some of the control methods may be incompatible with those diverse uses.

Generally, weeding using mechanical controls or by hand using various types of cutting, hoeing, and harvesting tools is the primary method of managing aquatic vegetation in many parts of the world (Wade, 1990). The secondary method is the use of chemical herbicides of which the predominant ones are 2,4-D, diquat, glyphosate, and various copper compounds (Murphy and Barrett, 1990). The use of chemical herbicides in water can result in residue and tolerance problems in potable water, nontarget effects when herbicide-containing waters are used for irrigation, and lack of selectivity.

These weeding and herbicide uses afford only temporary solutions; thus the cost of control by these methods is recurrent. Many public water bodies are intensively managed with respect to weed control, and the cost of such operations is often high. As a general rule weed management in public waters is underwritten by public funds, thus competing for the limited tax revenues.

Biological control can be a cost-effective, long-term solution (Andres and Bennett, 1975). Host-specific as well as nonspecific agents have been successfully used as biological control agents for several aquatic weeds. The herbivorous fish, the white amur or the grass carp (*Ctenopharyngodon idella*), has been used for a number of years in many countries for nonselective management of aquatic weeds, especially submerged weeds (Sutton, 1977). Among the weeds preferred by this fish are hydrilla, *Chara* spp., southern naiad, duckweed, and many other problem weeds. Since grass carp is not native to the Americas, a sterile triploid has been bred that cannot reproduce and become permanently established. Only this hybrid is allowed to be used for aquatic weed management in certain parts of the United States, and it is readily available from commercial suppliers in North America. When used at proper stocking rates, according to the size of water bodies and the nature of the weed problem, the grass carp offers an excellent, sustainable solution (Sutton and Vandiver, 1986).

Host-specific fungal pathogens also have excellent potential as biological control agents for aquatic weeds (Charudattan, 1990b). Several pathogens have been found that are capable of controlling aquatic weeds under experimental conditions. It has also been documented that *Cercospora rodmanii* and *C. piaropi*, two related species that cause foliar diseases of water hyacinth, cause natural epidemics capable of controlling this weed (Charudattan, 1986; Martyn, 1985). Similarly, natural epidemics have been associated with decline of large populations of aquatic plants in fresh water and brackish water systems (Charudattan et al., 1990). However, despite this potential, commercial development and use of microbial agents as bioherbicides to manage aquatic weeds has lagged because of the high cost of registering microbial herbicides under Federal Insecticide, Fungicide, and Rodenticide Act (FIFRA) requirements and because of the technological difficulties in producing practical levels of control under field conditions (Charudattan, 1990b).

Certain other miscellaneous agents such as snails (*Marisa cornuarientis* and *Pomacea australis*), the manatee (*Trichechus manatus*), the crayfish (*Orconectes causeyi*), and competing plants (e.g., *Eleocharis* spp.) have been considered as biological-control agents for aquatic weeds. However, practical use of these agents has not been realized because of various constraints.

Herbivorous arthropods have also had an unquestionable record of success as biological-control agents of aquatic weeds. Unlike the fish, weed-control insects are highly selective, and only host-specific agents are used. Outstanding examples of weed control by insects include management of alligator weed (*Alternanthera philoxeroides*) by the chrysomelid beetle *Agasicles hygrophila*), water hyacinth (*Eichhornia crassipes*) by the curculionid weevils *Neochetina eichhorniae* and *N. bruchi* and the pyralid moth *Sameodes albiguttalis*), and giant salvinia (*Salvinia molesta*) by the curculionid weevil *Cyrtobagous salviniae* (Harley and Forno, 1990).

Biological control of aquatic weeds has tremendous potential as a nonpolluting, ecologically sensible, cost-effective, long-term alternative to mechanical and chemical controls. Although some agents, such as the fish and insects, have gained considerable recognition as successful agents and have been used on a wide scale for many years, similar success can be obtained with other agents, notably microbial pathogens. Classical biological control against aquatic weeds using fungal pathogens is another promising approach (Charudattan, 1990c).

Despite the demonstrated success in controlling alligator weed, water hyacinth, and salvinia, up to now the overall importance of biocontrol has been relatively small compared to other control methods. Continued research and regulatory support will be key to gaining any future success and benefits from biological control.

Human and Environmental Health Concerns

It has been noted that pesticide use can have unintended adverse effects on beneficial organisms and on other nontarget species of plants and animals that come into contact with the chemical or its residues. However, the risks to human health posed by exposure to pesticides in the environment, drinking water, or food have become primary arguments in the debate about pesticide use.

Concern about chemical pesticides was brought to the attention of the public with the publication in 1962 of Rachel Carson's *Silent Spring*, a powerful and

---

Time Line for the Discontinuance of Four Major Broad-Spectrum Pesticides

Regulatory controls on pesticide use were established by the U.S. government to limit human exposure to residues and at the same time sustain the agriculture industry's ability to provide an abundant, nutritious, and safe food supply. Prior to 1972, approval of pesticide use was based on assurance that use would be safe and effective as sold in interstate commerce—the criteria by which most broad-spectrum pesticides were registered for use in the 1940s. After 1972, approval was based on tests that proved use would not generally cause unreasonable adverse effects to humans or the environment. Suspension of a pesticide use is based on a finding by the U.S. Environmental Protection Agency (EPA) that continued use would pose an imminent hazard. Unless cancellation proceedings are held, the use of the pesticide is reinstated. Cancellation of a use is based on findings that use will result in unreasonable adverse effects to humans or the environment. (Tolerances are set for residues of pesticides resulting from use on food. Revocation of a tolerance can also limit or cancel a use on food or feed crops.)

Below are listed four broad-spectrum pesticides that EPA removed from the market because they were found to have unacceptable impacts on nontarget species—including human beings. Decades lapsed between the time these pesticides were registered for use and the time of their suspension of use, cancellation of registration, or withdrawal from the market.

Chlordane—an insecticide used for termite control
December 1975:	EPA issues suspends registration for the use of chlordane
January 1977:	a Circuit Court of Appeals affirms chlordane use on termites
March 1978:	EPA cancellation hearings take place as registrant agrees to phase out uses of chlordane on corn and other crops
August 1987:	registrant agrees to stop sale for use on termites

EDB—a fumigant used to control plant pathogens
September 1983:	EPA issues a notice of intent to cancel registrations of EDB for major uses

---

poetic treatise about the unanticipated adverse effects of chlorinated hydrocarbons such as DDT. By 1962 several target pests were already resistant to DDT, a resurgence of secondary (minor) pests was associated with its use, and DDT began to accumulate in the food chain. DDT residues eventually were found in the blood and fat tissues and breast milk of humans (National Research Council, 1993b).

Partly in response to such concerns, the U.S. Environmental Protection Agency (EPA) was created in 1970 under the Nixon administration; authority for the regulation of pesticides was transferred from the U.S. Department of Agriculture to EPA. FIFRA was overhauled by Congress in 1972, providing EPA with new authority to regulate pesticides. EPA inherited a backlog of 40,000 registra-

Time Line for Discontinuance—continued

February 1984: EPA Administrator issues an emergency suspension

April 1984: EPA revokes tolerances for EDB residues for postharvest uses

DBCP—a fumigant used to control plant pathogens

September 1977: EPA issues a notice of intent to suspend registration of DBCP based on sterility studies of male industrial plant workers

November 1977: EPA issues a notice of intent to cancel registration for use on 9 food crops

July 1979: EPA issues a notice of intent to suspend all DBCP registrations after it is detected in groundwater, crop residues, and air samples from outside the pesticide application areas

October 1979: EPA issues a final decision to suspend all registrations except on pineapple fields in Hawaii

November 1979: EPA reinstates uses on pineapple fields in Hawaii

March 1981: EPA withdraws intent to cancel registrations for use on pineapple fields in Hawaii and accepts voluntary cancellations for all other uses

January 1984: EPA issues a notice of intent to cancel pineapple use in Hawaii

2,4,5-T—a phenoxy herbicide used to control broad-leaved weeds

February 1979: EPA issues an emergency suspension of all uses of 2,4,5-T

March 1981: a principal registrant negotiates a phase out of herbicide use

January 1985: EPA issues final decisions on suspended uses

February 1985: EPA issues final orders on nonsuspended uses

February 1986: No 2,4,5-T product is allowed for sale in the United States

tions for pesticides including some that had been in use for more than 25 years. Carson's legacy is that in 1972, EPA cancelled nearly all uses of DDT, most uses of aldrin and dieldrin in 1975, mercuric compounds in 1976, dibromochloropropane in 1977, and 2,4,5-T and Silvex in 1979.

Surface waters such as lakes and streams have been found to be contaminated by agricultural pesticides that run off the land as a result of chemical, soil, site, and climate conditions. Runoff is also associated with heavy early spring rains that may fall after herbicides are applied prior to planting. Agricultural pollution from runoff contributes to estuary decline and can harm nontarget organisms (National Research Council, 1993c). The presence of herbicides in surface water also seems to have significant adverse effects on fresh water phytoplankton and on zooplankton communities. The ecological implications of such population changes are poorly understood, but raise considerable concern about the effect agricultural pesticides have on the environment.

By the early 1980s, public concern had turned to the possibility of pesticide contamination of groundwater, a major source of drinking water for rural populations which was once thought not to be vulnerable to pesticides (U.S. Office of Technology Assessment, 1990). Scientific evidence indicates that groundwater contamination is due, in part, to agricultural pesticides that leach from above (Nielsen and Lee, 1987). In general, pesticides that are more readily soluble, less likely to be sorbed to soil particles, or more persistent (resist degradation) tend to leach into groundwater. In 1988, the U.S. Environmental Protection Agency detected 46 pesticides in groundwater from 26 states (U.S. Environmental Protection Agency, 1988), but was unable to conclude whether the source of contamination originated from normal or mismanaged uses of pesticides (U.S. Office of Technology Assessment, 1990). In addition, there is an incomplete understanding of the potential effects of groundwater contamination on health and environment. Nonetheless, experience indicates that cleanup of such contamination is difficult and costly (National Research Council, 1993a; U.S. Environmental Protection Agency, 1987).

Pesticides can be both acutely and chronically toxic to humans exposed to residues in the environment, drinking water, or food (National Research Council, 1986a, 1987, 1993b). Residues have been frequently detected in the soil, surface water, ground water and in foods (Baker and Richards, 1989; Hallberg, 1989; Holden et al., 1992; National Research Council, 1986a, 1987, 1993b; Thurman et al., 1991; U.S. Environmental Protection Agency, 1990).

The acute effects of a single exposure to high concentrations of pesticides have been documented from clinical and epidemiological studies (National Research Council, 1993b) and are of particular concern for those who apply pesticides. Some acute effects can be skin or eye injury, dermal sensitization, observable neurotoxic behavioral changes, or even mortality (National Research Council, 1993b). Acute effects also arising from consumption of contaminated foods are possible if application rates exceed legally mandated tolerances for pesticide residues or if residues of several pesticides with the same mode of action conjointly exceed those tolerances (National Research Council, 1993b).

The chronic effects of pesticides, caused by exposure to pesticides at concentrations below those causing acute effects, are more difficult to predict or detect in clinical or epidemiological studies. Certain tumors appear to be more common among farmers than in the population at large. This observation prompted a number of epidemiologic studies by federal and medical agencies. The incidence of cancer mortality in Midwestern farmers seems to be associated with corn production, according to studies conducted by the Iowa Institute of Agricultural Medicine and the National Cancer Institute. Cancer mortality in the southeastern United States has also shown a positive association with intensity of cotton and vegetable farming. The National Cancer Institute is studying the incidence of leukemia and non-Hodgkin's lymphoma among men from Minnesota, Iowa, and Nebraska. Increased rates of cancer have been the primary health concern related to long-term occupational and dietary exposure to pesticides, but neurological, immunological, reproductive, and developmental effects are now being included in chronic risk assessments (National Research Council, 1986a, 1993b). Most recently the potential for differential risks to infants and children from exposure to pesticides has been highlighted (Guzelian et al., 1992; National Research Council, 1993b).

Since the introduction and widespread use of broad-spectrum chemical pesticides, their chronic and acute risks to human and to environmental health have been documented. The public increasingly expresses preference for pest control systems that minimize these risks.

TIME TO REASSESS AND PLAN

To ensure the availability of safe, profitable, and durable solutions to pest problems in the next century, planning must begin now. The combined effects of resistance, escalating costs of developing new compounds, and pesticide-induced pest outbreaks seriously impede agriculture's ability to manage pests economically and safely using current broad-spectrum, chemical-dominated approaches. Because growers need nontoxic and low-cost controls for pest problems, researchers have been exploring a wide range of alternative management practices, including traditional cultural controls (Ferris, 1992). Reliable, safe, economic, and ecologically responsible solutions to pest problems are likely to be readily adopted by growers; therefore, the return on public or private investment in research directed at discovering new management strategies is likely to be substantial.

Planning for the future, however, requires a vision of the needs of society and of the scientific progress required to meet those needs. Chapter 2 describes that vision and strategies for implementing this new approach to pest management.

Defining and Implementing Ecologically Based Pest Management

There is a need to develop new pest-management systems that are long-term, cost-effective, solve unmet needs, and protect human and environmental health. As presented in Chapter 1, conventional, chemically based pest-management strategies encourage short-term solutions that can be harmful to the environment and to human health. Broad-spectrum chemicals also are ineffective against some pest problems. Future pest-management systems will be based on a broad knowledge of the agroecosystem and will seek to manage rather than eliminate pests. Agricultural practices that augment natural processes that suppress pests, where available, will replace existing practices that disrupt natural processes; and these practices will be supplemented with the judicious use of biological-control organisms and products, target-specific chemical pesticides, and pest-resistant plants. It will also be necessary to reopen and develop channels of communication at all levels that increase the flow of information and cooperative action, subsequently lowering risk to users, fostering interdisciplinary interaction, and improving the profitability of alternative pest-control methods.

GOALS OF ECOLOGICALLY BASED PEST MANAGEMENT

The fundamental goals of EBPM are (1) safety, (2) profitability, and (3) durability.

• Pest-management systems must be safe for growers and workers who use them, and for consumers of the food produced. Minimizing health risks must be a primary criterion of acceptability of novel management systems.

• Pest control strategies must be cost-effective as well as effective, easy to implement, and readily integrated with other crop-production practices. Economic factors involving cost-effectiveness include crop prices, costs and availability of labor, land, equipment, and other production inputs.

• Pest-management programs must ensure that pests in the agroecosystem can be managed over the long-term without adverse environmental, economic, or safety consequences. Current pest-management strategies that rely on repeated applications of conventional broad-spectrum pesticides encourage the development of resistant species. In new ecologically based approaches, addressing potential development of pest resistance will be important.

EBPM promotes the economic and environmental viability of agriculture by using knowledge of interactions between crops, pests, and naturally occurring pest-control organisms to modify cropping systems in ways that reduce damage associated with pests. Ecologically based management relies on a comprehensive knowledge of the ecosystem, including the natural biological interactions that suppress pest populations. It is based on the recognition that many conventional agricultural practices disrupt natural processes that suppress pests. Agricultural practices recommended by EBPM will augment natural processes, supplemented by biological-control organisms and products, resistant plants, and targeted pesticides.

An ecosystem is dynamic with interacting physical, chemical, and biological processes. The coexisting crops, herbivores, predators, pathogens, weeds, and other organisms interact with one another and respond to their environment. Each organism has developed a repertoire of offensive and defensive maneuvers in response to changes in the behavior of other organisms in the cropping system. This web of interrelated interactions also confers stability on the system; while a population increases and decreases, it is subject to the checks and balances imposed by populations of the other organisms.

Stability (i.e., low variance in density of pests over time) is an essential feature of successful pest management. When effective predators, parasites, pathogens, or competitors of potentially destructive pests are present in the managed ecosystem, pest populations are suppressed and held in check. In natural systems, biological-control organisms are often quite diverse, leading to stable, low pest populations.

Activity of most biological organisms is density dependent—i.e., when pest density is low, density of, and hence suppressive activity of control organisms tends to be low, and vice versa. Negative feedback related to population density keeps both pest and control organism from both glut and extinction. Because neither pesticides nor host-plant resistance methods are responsive to feedback, achieving stability and balance within the agroecosystem is not possible with those systems, but is a fundamental goal of EBPM. EBPM is founded on the importance of natural processes inherent within agricultural and forest production systems. To this is added, in a complementary way, other technologies for managing pest problems.

Ecological balance is more difficult to attain in a highly modified agricultural environment, such as a large-scale monoculture farm, where the goal is to maximize production of one crop species exclusively. In this monoculture ecosystem, biological-control organisms that depend on other plant species for growth and reproduction may suffer tremendous population reductions. On the other hand, a pest adapted to utilizing the primary crop has an unlimited resource at hand, resulting in a pest population explosion.

The National Research Council report on alternative agriculture describes alternative systems to manage pests, crop nutrients, soil erosion, and livestock production (National Research Council, 1989b). The goal of alternative agriculture—profitable, safe, and healthy ecosystems achieved by integrating individual practices into an overall farm management system—are consistent with the goals of EBPM.

EBPM should be viewed in the context of whole-farming systems. Pest-management methods cannot be isolated from other components of agronomic systems such as fertilization, cultivation, cropping patterns, and farm economics (National Research Council, 1989b, 1991). These physical, biological, and chemical practices are interrelated; changing one system component will impact another entity. For example, choosing a particular rotating crop can augment suppression of soil-borne plant pathogens and affect levels of soil nitrogen. Such natural processes of interdependence are augmented and exploited by ecologically based pest-management systems. Biological and ecological processes are fundamental to pest control even in the most intensively managed ecosystems; EBPM builds on and supplements them, rather than impeding or replacing them.

SUPPLEMENTS TO NATURAL PROCESSES

A major premise of EBPM is that most potential pest species are held in check by naturally occurring beneficial organisms. Supplemental inputs, either natural or synthetic, must not suppress either the populations or activities of these indigenous beneficial organisms. It is therefore essential that the use of supplemental inputs be based on an understanding of target organisms so that the potential for development of resistance, disruption of natural and biological processes of control, and unintended effects on nontarget organisms or ecosystems are minimized. Supplemental inputs that meet the criteria of safety, profitability, and durability are valuable and, quite possibly, finite resources. Their use should be accompanied by sophisticated diagnostic and monitoring tools and methods of deployment that prolong their effect. Lasting solutions require anticipation of potential disruptions and evolutionary responses that could result from pest-management practices (Gould, 1991).

The self-sustaining system requires no supplemental inputs, relying instead

Integrating Components of a Managed Ecosystem: Cover Crops

Cover crops are an excellent example of an innovative strategy to integrate multiple components of a managed ecosystem. These noncrop plant species, such as vetch and clover, are grown as ground cover to manage pests, provide nitrogen for subsequent crops, increase soil organic matter, and reduce soil erosion (National Research Council, 1989b). Because cover crops increase ecosystem biodiversity which, in turn, affects multiple biological interactions involving pest management, soil fertility, and plant nutrition, ecosystem interactions should be carefully considered when integrating a cover crop into a pest-management strategy. Since the long-term impacts of cover crops are not well known, additional research can provide a greater understanding of their role in crop production (Hanna et al., 1995).

Pests

Cover crops can provide habitat and a food source for biological-control organisms. California vineyard managers plant clover and other legume ground covers to attract beneficial wasps and spiders; their abundance is associated with decreases in leafhopper pests (Hanna et al., 1995; National Research Council, 1989a). Some cover crops produce allelopathic compounds that can suppress plant parasitic nematodes. Ground covers also can delay weed emergence, giving a competitive edge to the primary crop.

However, interactions of a cover crop with other ecosystem components can lead to undesirable effects; a cover crop species that is optimal for biological control of arthropod pests may not be competitive with a troublesome weed (Ingels, 1995). Thus, considerable knowledge of ecosystem processes is necessary to successfully manage pests in cover-cropping systems.

Soil Fertility and Plant Nutrition

Cover crops can increase soil fertility and plant nutrition; legumes such as vetch and clovers exhibit powerful nitrogen-supplying capabilities. Symbiotic bacteria initiate nodules on the roots of legumes, which transform atmospheric nitrogen into a useful nitrogen source for plant growth (nitrogen fixation). This complex nitrogen-transformation process is influenced by numerous factors including soil microorganisms, cover crop species, tillage, and water (National Research Council, 1989b).

Since cover crops may also remove nutrients from the soil, nutrient status of a primary crop needs to be monitored. Thus, growers need to weigh the benefits and disadvantages of using cover crops to increase soil fertility and plant nutrition.

on natural, biological processes. Supplemental inputs will undoubtedly be required, however, to achieve EBPM. These will include biological-control organisms and products, narrow-spectrum synthetic pesticides, and resistant plants.

Biological-Control Organisms

Biological-control organisms are living organisms that can be used to manage arthropod (mites and insects), weed, and plant (bacteria, fungi, viruses, and nematodes) pests and pathogens. Generally, control organisms do not immediately alleviate disease or prevent or curtail attack; they also do not immediately reduce the pest population. Rather, control is generally achieved over a period of several generations. Biological control organisms commonly interacting with their hosts at low population densities, preventing pests from reproducing to economically important population levels. Control organisms are themselves arthropods, plants, and pathogens and are as diverse as the pests (Ferris, 1992; Flint, 1992; Schroth et al., 1992; Turner, 1992):

- arthropods that prey on or parasitize other arthropods,
- arthropods that prey on or parasitize plants,
- pathogens of plant pests,
- bacterial or fungal antagonists of plant pathogens,
- beneficial nematodes that parasitize arthropods,
- mild strains of plant pathogens, and
- other beneficial organisms that parasitize or prey on plant pathogens or nematodes.

Predatory arthropods can be very host specific, depending on their ability to locate, consume, and utilize a particular prey species for growth and reproduction; however, environmental factors and habitat may modify prey specificity. A predator invariably consumes more than one individual during its life span and, if conditions are favorable, some arthropod organisms kill hundreds of host individuals during their development. Arthropod parasitoids attack and disarm the arthropod host species and subsequently deposit one or more eggs within or on the host organism. Parasitoid larvae then feed on and complete development on a host individual, and in the process, kill the host. Parasitic organisms are usually highly host and habitat specific. Arthropod herbivores that prefer feeding on weed plants can be used as controls, feeding on foliage, roots, stems, flowers, fruit, or seeds of weeds.

Viruses, bacteria, fungi, protozoa, and other microbes that cause disease in arthropods, plant pathogens, or weeds are also used as biological-control organisms. Under favorable conditions, they infect their hosts and can cause epidemics that can lead to a marked decline in the pest population. Certain of these persist on the plant, in the pest, or in the environment, causing recurring infections in their hosts.

Microbial antagonists can suppress plant pathogens by producing antibiotics or by means of competition—producing larger populations, and thus occupying and competing for the same ecological niche. A competitor will challenge the pathogen for infection sites, nourishment, or other resources and, in the process, reduce the population size of the plant pathogen. Microbial antagonists can produce toxins active against arthropods, plant pathogens, or weeds, or compete with plant pathogens for nutrients or preferred sites on plant surfaces. Mild strains of plant pathogens that cause little or no disease can induce resistance responses in the plant or otherwise provide protection from disease.

Biological-Control Products

Genes or gene products derived from living organisms that kill, disable, or otherwise regulate the behavior of plant pests are biological-control products. Examples include the *Bacillus thuringiensis* (Bt) toxin, delivered by a killed microbe, and pheromones or other semiochemicals used to kill or disrupt the reproduction of arthropod pests.

It must be noted that genes or gene products are derived from living organisms, but they are not inherently more suitable supplements than synthetic products. Indeed, certain synthetic products can be less detrimental to environmental balances than some products derived from organisms; some natural plant products used to formulate botanical insecticides such as rotenone and pyrethrum have broad-spectrum activity and can be highly toxic to beneficial organisms. It is the spectrum and activity of products used in ecologically based pest management that are of primary importance rather than the source of the products. The most useful biological-control products are those that have minimal impact on all components of the agroecosystem—except for the target pest.

Synthetic Chemicals

Narrow-spectrum synthetic pesticides that meet the criteria of safety, profitability, and durability are suitable supplements for EBPM. For example, the synthetic insecticide pirimicarb is highly selective for one group of pest arthropods, aphids, and does not adversely affect most biological-control organisms.

Resistant Plants

Plants that have developed resistance against pests will be important components of EBPM. Plant breeders have successfully identified and deployed genes for disease and arthropod resistance in numerous crops; in the future, molecular genetic methods will become more important as a means to producing pest-resistant plants. At present, resistance is the predominant defense against many plant diseases, such as rust diseases, that would otherwise severely limit cereal

Pirimicarb: The Saga of a Selective Pesticide

The insecticide pirimicarb is a selective aphicide; it controls many of the most damaging aphid pests but has little if any effect on most other arthropods. Because it does not directly interfere with beneficial predatory and parasitic arthropods, it fits well in alternative pest-management programs that rely on biological control. However, pirimicarb had a relatively short early history in the United States; it was first registered in 1974 but was voluntarily withdrawn from the market in 1981 because of regulatory and marketing problems.

Pirimicarb was registered only on specialty crops, specifically potato and greenhouse crops, where aphids are serious pests. After initial registration, the Environmental Protection Agency (EPA) requested additional metabolism and residue information that would have been very expensive to gather. Also during this time, the synthetic pyrethroid insecticides were coming on the market, and many pest managers preferred products such as these that had broad-spectrum activity. Faced with the economics of clearing regulatory hurdles as well as facing competition from the new pyrethroids, it was decided to withdraw registration of pirimicarb in the United States, even though the product continued to be used in Canada and Europe.

Over the past 20 years, the climate in the United States has improved to favor the use of selective products that provide effective and economic alternatives to more broad-spectrum pesticides. Pirimicarb is currently undergoing reregistration review in Europe. Data required for this review will satisfy some of the EPA requirements; therefore the parent company intends to once again submit pirimicarb for registration in the United States.

Had a mechanism been in place in the 1970s to foster the development of specific pesticides by helping companies meet regulatory requirements, pirimicarb and other selective pesticides would likely be more widely used in agriculture today.

SOURCE: Personal communication, 1995, M. Moss, ZENECA Ag Products, Wilmington, Delaware.

crop production in much of the world. In the case of rust diseases, a plant cultivar generally is resistant to only one specific race of a pathogen. Other races of the pathogen can infect the plant, and the shift in race composition of the pathogen leads to a boom-and-bust syndrome of rust diseases. Strategies of resistance-gene deployment, in which fields are planted with mixtures of cultivars, each with a different race-specific resistance gene or with one cultivar containing multiple race-specific resistance genes, can be very successful in diminishing this syndrome. Race-specific resistance genes deployed in this manner can be quite successful in controlling plant diseases. Plants also may have a general resistance to plant pests, conferred by the collective action of multiple genes. Polygenic resistance can be quite stable and for this reason will be an important component of EBPM.

To date, resistant plants have been developed almost entirely through plant breeding. Future breeding programs will continue to rely on diverse wild germplasm as a source of resistance genes but will also incorporate resistance genes identified in research programs. These investigations will enhance the plant's inherent strength to survive in its environment. There is good reason to believe, based on the tremendous recent progress in identifying pest-resistance genes, that numerous genes identified by these approaches will be incorporated into crop plants in the future. The promise of durable resistance can only be reached, however, if breeding programs also strive to enhance, rather than diminish, the genetic diversity of plants grown in forest or agricultural ecosystems. Stable and long-lasting pest management will depend on the availability of crop plants with broad bases of genetic variability.

ECONOMIC FEASIBILITY OF ECOLOGICALLY BASED PEST MANAGEMENT

EBPM will be implemented on a farm level and must be profitable for the grower. Adoption of an alternative pest-management strategy depends on its relative profitability, risk, public policies, and the information and education available to the grower. The realistic potential of EBPM systems will, in large part, depend on how feasible those systems appear to the individuals who must implement the systems. Management systems that effectively suppress pest populations but suffer from poor profits, high risks, discouragement by public policy, or lack of available information for the grower will not be implemented.

The economic feasibility of pest management must be determined by examining the economic factors a grower might consider when considering adoption of EBPM strategies. It should be noted that a larger knowledge base is necessary to make economic comparisons of EBPM strategies.

Economic Feasibility of Pest Management

Economic feasibility, as defined by Reichelderfer (1981), refers to the likelihood that a management system will bring net returns greater or equal to that of any other management system being considered by a grower. A grower will be encouraged to adopt an ecologically based technology if it results in net profit at least as great as does the system the grower is currently using. Relative profit is a great incentive to adoption. Indeed, if the profit margin is great enough, growers may even be induced to alter their management styles in order to take advantage of the new opportunity.

Economic feasibility does not consider social costs and benefits, but it is the starting point for broader analysis of the desirability of a pest-management sys-

Biological Control of Citrus Pests

Worldwide, efforts to develop ecologically based approaches to arthropod, pathogen, and weed control for citrus production are producing diverse, effective, and economical alternatives to frequent, heavy applications of pesticides. Some of the methods noted below are well established, some are being rediscovered, and others are still in developmental stages.

Biological Control of Arthropods

Augmentation of Existing Control Organisms

In California, the California red scale (*Aonidiella aurantii*), one of the primary citrus pests, is now controlled by augmenting populations of *Aphytis lingnanensis*—a parasitoid of the red scale. Normally *A. lingnanensis* is more abundant in the summer, whereas adult female red scales accumulate in the spring. The control method is to release the parasitoid, commercially raised in grower-owned cooperatives, in the spring when the adult female red scale appears. The parasitoid is active against the female red scale before it can reproduce, thus eliminating the need for multiple applications of broad-spectrum scalicides (Graebner et al., 1984).

Commercially produced microbial pesticides that have been investigated for use in the citrus system include the fungus *Hirsutella thompsonii*, which is active against the citrus rust mite *Phyllocoptruta oleivora*, the most important citrus pest worldwide. Although problems with formulation of this product (Mycar®) limited the availability of this fungus in the 1980s, researchers continue to seek methods to solve these difficulties (McCoy and Couch, 1982). A commercial product currently in use for biological control of root weevils in Florida citrus is the entomophagous nematode *Steinernema riobravis*. This beneficial, soil-inhabiting nematode is cosmopolitan in distribution, but occurs naturally in soils at low levels, insufficient for effective management of the weevil larvae and pupae. Commercial fermentation culture has led to the marketing of a product (Biovector®) that is applied to the soil beneath citrus trees. The degree of success with this biological-control product remains to be determined, as the product has been in use for only 3 years (McCoy and Duncan, 1995). An acaricide (miticide) recently registered for use in citrus, Avermectin®, interferes with molting and transformation of mites from nymphs to adults. This growth regulator is an example of biologically based products that can be employed in pest management (Knapp, 1995).

Conservation of Existing Control Organisms

The importance of already existing natural processes of control is illustrated by the example of the bayberry whitefly, *Parabemisia myricae*, an introduced pest from Japan that became established in California citrus groves. Parasitoids of the whitefly were found in Japan, and several species were introduced into California, but without successful control. However, in 1982 populations of the whitefly de-

Biological Control of Citrus Pests—Continued

clined dramatically. It soon became apparent that a native parasitoid *Eretmocerus debachi* had begun to attack the bayberry whitefly and eventually reduced it from a serious pest to simply another of the many innocuous species that inhabit citrus groves.

Classical Biological Control

The most famous use of an exotic biological-control organism to achieve permanent control of an arthropod pest of exotic origin is the control of the cottony-cushion scale, *Icerya purchasi*, that threatened the continued existence of the California citrus industry in the late 1800s. The predaceous Vedalia beetle, *Rodolia cardinalis*, was introduced into California citrus groves in 1889. By 1890, the cottony-cushion scale was no longer a threat to California citrus production (DeBach and Rosen, 1991). This success in California led to similar successful introductions of the Vedalia beetle into citrus groves in Florida, Texas, and eventually worldwide in 25 other countries (DeBach, 1974).

Biological Control of Plant Disease

Cross-Protection against Citrus Tristeza

Cross-protection against citrus tristeza virus has been used for more than 2 decades in Brazil, where millions of citrus trees have been protected against highly virulent strains of the virus by mild strains (Costa and Müller, 1980). The citrus tristeza virus is a serious disease organism that has crippled commercial industries worldwide. Transmitted by aphid vectors, citrus tristeza virus has escaped attempts at management through aphid control, disease therapy, and continued efforts to develop effective host-plant resistance. Inoculation of healthy susceptible trees with a harmless "mild strain" of the virus has been demonstrated to confer protection against subsequent inoculation with more virulent strains.

Suppression of Postharvest Rot

Early in 1995, two microorganisms were registered by the Environmental Protection Agency for the biological control of postharvest diseases of citrus. The yeast *Candida oliophila* was registered for control of postharvest rot of citrus and apple, and the bacterium *Pseudomonas syringae* was registered for control of storage rots of citrus, apple, and pear (Wilson and Janisiewicz, 1995).

Biological Control of Weeds

A host-specific strain of the pathogenic fungus *Phytophthora palmivora* has been developed as the commercial mycoherbicide DeVine® to manage milkweed vine (*Morrenia odorata*), which infests citrus groves.

tem (Headly, 1985; Reichelderfer, 1985). Reichelderfer (1981) and Carlson (1988) identified several factors which determine economic feasibility. These factors can be grouped either as pest control factors or economic factors. The interrelationships between these two factors indicate how difficult it is to achieve economic feasibility, and the need for biological and social scientists to cooperate in research, development, and distribution of ecologically based management systems (Headly, 1985; Reichelderfer, 1981).

Pest-Control Factors

Three pest-control factors that have important effects on the economic feasibility of a pest-management system are (1) the severity of pest-induced losses, (2) the variability of pest populations, and (3) the technical efficacy of the management system. These factors are defined in the discussion below.

A pest-management approach is economically feasible only if it reduces an important pest population to an extent that it no longer limits profitability. If even low population densities of the pest can cause serious damage, economic losses may remain too high after the management system is implemented and the relative economic benefits from the management system will decrease (Carlson, 1988; Reichelderfer, 1981). Hence, the economic feasibility of an ecologically based pest-management system will depend on how much damage the steady-state population of the pest can cause. Economically feasible, ecologically based control may be easier to effect for pests that can be tolerated at moderate populations without economic damage.

The variability of the pest population over time and space can also affect economic feasibility of a management system. If the pest is only a problem every few years, then an economically rational decision would be to wait until the pest surpassed an economic threshold before treating it. Waiting to employ a biological-control organism until the pest problem becomes severe, however, may not be feasible for all biological-control organisms and depends on their method of deployment. Inoculating the crop to prevent a pest problem that occurs only infrequently will result in unnecessary expenses in some or perhaps most years. In cases where the pest problem occurs frequently and predictably, routine inoculation with biological-control organisms may be economically feasible (Carlson, 1988; Reichelderfer, 1981). In contrast, some augmentative or classical biological control can provide permanent population reduction below economic levels, i.e., prevent the pest from becoming a pest. This is an advantage in many cases—classical biological control using organisms is preventative rather than therapeutic.

Technical efficacy of an ecologically based management system refers to the ability of the system to prevent or reduce damage caused by pest populations. As the management system becomes more efficacious, it becomes more economically feasible, assuming there is no change in other factors such as the cost of other production inputs, the prices of commodities, or the effectiveness of alternative management systems (Reichelderfer, 1981).

Economic Factors

Economic factors such as crop price and yield, costs of alternate management methods, and implementation costs determine the economic return a grower realizes from use of an alternative pest control system.

Crop price and yield determine the gross return a grower receives per hectare. As the gross return goes up, so does the value per unit of pest damage, which in turn increases the value of management systems that decrease the damage (Carlson, 1988; Reichelderfer, 1981). Crops that produce large gross returns per hectare create strong incentives to invest in pest-management systems. Growers will, in general, be more willing to invest in alternative pest control systems when the value per hectare of the crop they are producing is large. Ecologically based management systems that solve pest problems that cannot be solved with current systems will be immediately attractive to growers, particularly if the cost of current pest damage is high.

New, ecologically based management strategies will compete for adoption and implementation with the systems currently used by growers. In most cases, the current system is based on the use of broad-spectrum, conventional pesticides to kill pests. The costs of an alternative management system include the costs of (1) pesticide, biological-control organism, biological-control product, or other supplemental input, (2) capital investment of machinery, (3) machinery operation, (4) management time, and (5) labor (Carlson, 1988).

Some ecologically based methods are much cheaper than current systems. Reichelderfer (1981) and Cate and Hinkle (1993) for example, noted the cost advantages of classical biological-control methods over pesticides. The benefits of permanent, successful pest management achieved with a one-time introduction of a biological-control organism will, over time, be more cost-effective than annual pesticide applications. Costs of seasonal releases of a biological-control organism may be competitive with pesticidal alternatives (Reichelderfer, 1985).

Implementation costs are the costs imposed by switching to and learning a new management system. These costs also include the time and money the grower must invest in physical and human capital to be able to use the new management system. Switching to a new system may require training or the purchase of equipment and retirement of current equipment. Initially, more labor and presumably more management for an indefinite time period would be required to learn and integrate the management system into the farming system. Any effects of a new pest-management system on farm program eligibility, off-farm employment, or other factors affecting income are also included in implementation costs.

Implementation costs are very important considerations for growers deciding to implement alternative pest-management systems. The more similar the new management system is to the current system, the less likely new machinery will be needed, and the easier it will be to learn the new system—all resulting in low implementation costs. Biological-control organisms formulated as seed treatments, for example, may be readily integrated into current agricultural practices and require no specialized equipment for implementation.

Plant breeding for arthropod and disease resistance is an excellent example of an ecologically based approach that has been easily integrated into current production systems. Growers regularly substitute one resistant variety for another with very low implementation costs. Indeed, the ongoing development of new resistant varieties is the primary line of defense against important plant diseases such as wheat rusts that annually spread into the United States from Mexico.

The potential of ecologically based management systems may depend as much on how well they meet the economic criteria of those who use them, as on their direct effect on pest populations.

Economic Feasibility and Risk

Risk plays a large role in a grower's decision to adopt a new pest-management system. In the case of growers,

> the value of controlling a pest, whose incidence varies in an uncertain way, is greater than the average loss caused to them by the pest. Risk-averse growers are willing to pay a premium or insurance charge to reduce the risk of uncertainty they face (Tisdell et al., 1984:p. 172).

In other words, growers are interested in minimizing the variability surrounding returns and yields as well as the absolute return or yield achieved (Headly, 1985).

There are tradeoffs between expected net returns and year-to-year variability of returns (Kramer et al., 1983; Reichelderfer, 1981). Kramer et al. (1983) found that profit maximization and risk aversion were the most important criteria determining the choice of soil conservation technologies. Even when required to meet reduction standards, risk-averse growers chose the more erosive systems to guarantee less net return variability; risk-neutral growers adopted more practices optimizing soil conservation.

Growers have been quick to adopt pesticides because of the certainty pesticides bring to production and profitability, and increased uncertainty about pest damage results in increased use of pesticides. Too much variability in net returns from year to year may induce a risk-averse grower to select a pest-management system that produces more certain results even if average returns are lower (Headly, 1985; Tisdell et al., 1984). Risk-averse growers with tight cash flows may decide to use a production system that brings a lower average return but has less income variability than a production system that is more variable but has a higher average return. This also could speak to technology systems where the returns over the life of the investment are high but initial returns are low.

Uncertainty about the efficacy of ecologically based management systems is a major source of concern for growers. Uncertainty surrounding the effectiveness and consistency of alternative pest-management systems has been a barrier to the adoption of IPM practices. Fernandez-Cornejo et al. (1992), for example, found that adopters of IPM technologies were less risk-averse than those who did not adopt IPM. Risk-averse growers were more likely to rely on a conventional pesticide system (Fernandez-Cornejo et al., 1992) and placed more value on immediate reduction of the pest (Reichelderfer, 1985). Risk averse growers may be reluctant to forgo application of a pesticide while waiting for uncertain results from an ecologically based system.

Risk also depends on the stability and supply of the biological-control organisms, biological-control products, or other supplemental inputs to ecologically based systems. A steady supply of biological-control organisms or other inputs to the system or knowledge is essential to insure that growers can find what they need, when they need it. Though supply shortages can also occur in conventional pest-management systems, uncertainty about the availability or use of an essential component of an ecologically based system increases the risk to the grower of using that system.

The interaction of economic feasibility and risk largely determines the likelihood that an ecologically based management system will be adopted or implemented by growers. Economic feasibility and risk can create both barriers to or opportunities for the implementation of ecologically based management systems. A national initiative to develop and implement ecologically based systems should focus on those strategic opportunities where the economic feasibility and risk characteristics increase the likelihood of eventual adoption and implementation by growers.

Safety, profitability, and durability are not mutually exclusive, but the public interest in reducing risk to human and environmental health may outweigh private considerations of economic feasibility. Directing investments toward ecologically based systems that are economically feasible and less risky for growers will help ensure that the systems are profitable and at least economically durable. Growers, however, cannot consider all the social and environmental costs of alternative management systems when they make their decisions. For instance, society needs to consider costs to monitor for potential development of resistance of EBPM solutions implemented in managed agricultural and forest ecosystems. The public at large also benefits from effective, long-term solutions to pest problems that minimize environmental and health risks. Strategic opportunities to encourage grower adoption and minimize the ecological and human health risks should also be explicitly identified as part of the planning for a national initiative to implement ecologically based management systems.

As a first step, an ecologically based management initiative should be directed to systems that *initially*

- promise to reduce risks to human or environmental health posed by current management practices,
- promise to achieve lasting solutions to pest problems,
- solve pest problems that have no feasible pesticide solution,
- are less expensive than conventional pesticides or are applicable to high-value crops,
- require minimal changes in current production systems, and
- promise to reduce the risk and variability of annual returns to the grower.

Ecologically based systems that meet one of these criteria are targets for investment of public resources; systems that meet several of these investment criteria are the most promising targets.

Growers are likely to adopt EBPM systems that generate lower risks and higher profits. There is always perceived risk in embracing new technologies; the greater the divergence from previous practices, the greater the perceived risk. The value of information is that it reduces the pest manager's uncertainty about pest control decision making (Lawson, 1982), thereby increasing the likelihood of acceptance of a new practice. The public at large has an opportunity to invest in new knowledge and tools that will help the grower successfully implement ecologically based pest-management systems.

THE ROLE OF INFORMATION IN PEST MANAGEMENT

The complexity of managed ecosystems indicates a need for more multi-disciplinary information to develop and implement EBPM. All technological advances are driven by information flow; the importance of this for development and adoption of new pest-management technologies has been well documented (e.g., Edwards and Ford, 1992; Fitzner, 1993; Frisbie, 1989; Frisbie et al., 1992; Grieshop et al., 1988; Lawson, 1982; Mumford, 1982; Norton, 1982; Poston, 1989; Putter and Van der Graaff, 1989; Scott and Harris, 1989; Tette and Jacobsen, 1992; Van Driesche, 1991; van Lenteren, 1989; Zalom and Fry, 1992). Conventional pest management itself requires a high level of information, some of which is not being effectively transferred from research to the field. Hoy (1989) stated that "... ineffective transfer of information from the developmental phase to the implementation phase seems to explain why new pest-management techniques are not implemented more frequently." Ecologically based IPM will require an even higher level of knowledge to guide management decisions (Edwards and Ford, 1992; Fitzner, 1993). Grieshop et al. (1988) stated that "attributes of [an] IPM innovation, particularly its complexity, figure prominently in adoption rates." Tremendous complexity of EBPM may result in slow adoption rates. Therefore, we are faced with a dilemma; current IPM technology

Integrated Management of Frost Injury and Fire Blight

Frost Injury

Frost injury is a serious problem on agricultural plants, most of which cannot tolerate ice formation within their tissues. Annual losses to agricultural production in the United States alone are estimated at more than $1 billion. At temperatures only slightly below freezing ($-10°$ C and above), even frost-sensitive plants have a natural ability to super cool, thereby avoiding ice formation within their tissues. The capacity to super cool is limited, however, by the presence of ice-nucleation active (Ina$^+$) bacteria, especially *Pseudomonas syringae*, a common inhabitant of aerial plant surfaces. Ina$^+$ bacteria have an outer-membrane protein—the ice protein—that orients water molecules in an arrangement that mimics the crystalline structure of ice. The oriented water molecules do not super cool but, instead, freeze at temperatures very close to freezing (i.e., $-2°$ C to $-5°$ C). Once ice formation is catalyzed by Ina$^+$ bacteria on the plant surface, ice crystals rapidly spread into the plant, injuring plant tissues irreversibly. Injury at temperatures close to freezing (i.e., $-2°$ C to $-5°$ C) can be avoided by reducing the number of Ina$^+$ bacteria.

Fire Blight

Fire blight, an important disease of pear and apple, is caused by the bacterium *Erwinia amylovora*. The pathogen can grow on shoots, leaves, or blossoms; enter the plant; and then can grow internally, sometimes killing the tree. At present, growers manage fire blight by spraying trees with the antibiotics streptomycin or Terramycin, which reduce the population size of *E. amylovora* on the plant surface. Resistance of *E. amylovora* to streptomycin, which previously was the most effective antibiotic, is now common in many pear-growing regions of the United States.

Biological Control

Frost injury on a variety of agricultural plants and fire blight of pear and apple can be suppressed by BlightBan®, a product composed of the bacterial-control agent *Pseudomonas fluorescens* A506. After even a single application of Blight-Ban, large populations of *P. fluorescens* A506 can develop on treated leaves or blossoms. These populations compete with *E. amylovora* or Ina$^+$ *P. syringae* for limiting nutrients on plant surfaces. The severity of fire blight and frost injury is reduced because the population size of the causal organisms is reduced through competition with the beneficial bacterium. Although A506 has been as effective as streptomycin in suppression of fire blight in many field experiments, it can also be used in concert with conventional practices for management of fire blight. It is naturally resistant to streptomycin and Terramycin, so can be combined with antibiotics in spray programs. The opportunity to integrate chemical and biological control is attractive to growers, who are more likely to adopt biological control initially if risks can be minimized by combining it with familiar and effective methods.

SOURCE: Lindow, S. E. 1985. Integrated control and role of antibiosis in biological control of fire blight and frost injury. In Biological Control on the Phylloplane, C. Windels and S. E. Lindow, eds. St. Paul, Minn.: APS Press.

is sometimes transferred at a slow pace and EBPM will require different and even more information. An intense and well-coordinated information, education, and training initiative is going to be essential to resolve this dilemma and move agriculture from conventional, chemically based pest control to EBPM.

Putter and Van der Graaff (1989) identified four levels of decision makers in the pest-management process: growers, research and extension workers, regulators, and the private sector. Here we expand this list to the following six functional groups: (1) researchers, (2) educators, (3) policy makers and regulators, (4) private business, (5) the general public, and (6) end users. The information each of these groups needs may be different, but facilitating the flow and use of information to and within each group is essential. A breakdown in information flow to one of the six groups may seriously hamper the development and implementation of EBPM, as it has in the development of other technologies. Although the timely and unrestricted transfer of new information between all of these groups is critical, here we will emphasize only the importance of this process to the end users (Figure 2-1).

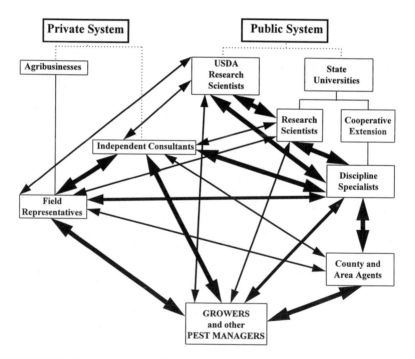

FIGURE 2-1 The information web of ecologically based pest management. The intricate and complicated structure of this graphic is intentional, meant to provide a visual presentation of the complexity of the pest-management information web. The actual flow of information shown above indicates the difficulties in transferring information to end users.

End users are those who make and carry out pest-management decisions. They include growers, crop consultants, pest control advisors, corporate field representatives, nursery and greenhouse managers, landscape maintenance professionals, forest managers, stock handlers, stored product managers, pest control operators, gardeners, and others. These end users are referred to collectively as pest managers. Delivery of information to this group is crucial to changing from current chemically based pest management to ecologically based strategies. Regardless of the technologies available, if end users do not have a clear understanding of the use and impact of these technologies, change will not occur. Although this is the most important link in the pest-management chain of knowledge from basic research to implementation, the total linkage of information transfer to the end user needs to be strengthened and streamlined.

Putter and Van der Graff (1989) have indicated that pest managers are decision makers, and that sound decisions require sound information. Knowledge, or at least ready access to appropriate information, is essential to successfully and economically manage pests. Efficient information flow is critical to successful pest management because (1) the knowledge needed is complex and a large amount of information is available; (2) there is a diversity of sources of information, both public and private; and (3) pest situations are dynamic and change substantially in both the short term and the long term (Lawson, 1982).

Putter and Van der Graff (1989) stated that growers must be knowledgeable in three areas to successfully manage pests:

1. they must have information about the pests, including biological characteristics and type and significance of damage;

2. they must have information about pest control options and the implications of using these options; and

3. they must be knowledgeable about the economic implications of specific pest-management decisions.

Norton (1982) proposed a general approach that can be used to improve crop protection decision making. This approach can be viewed as closing "information gaps" and consists of a three-step process:

1. identifying the types of information needed by the pest manager,

2. identifying how the information is needed, and

3. analyzing how the information gets to the grower.

Norton (1982) distinguished between "total need" and "specific need" information. To satisfy the total need, the pest manager must be familiar with all options and thus require a breadth of information. For example, the pest manager must know the entire range of cultural controls, biological controls, chemical controls, and host resistance options available. To satisfy the specific need, the pest manager must know detailed information about each of the individual practices being considered. For example, if releasing a natural enemy for augmentative biological control, the pest manager must know the proper time of release based upon susceptible stages of the target pest(s), rates of release based upon pest pressures and efficacy under different environmental conditions, range of target pests controlled, and cost.

Information has to be provided to the end user at the appropriate technical level, and in a clear and convincing presentation. Both the information and the delivery can take many forms. Although we often think of the formal channels of communication (such as through extension programs, publications, trade journals, or industry advertising), informal channels also play a key role. The innovative end user will be the first to evaluate and adopt new technologies; the actions taken by this person reduce the risk perceived in the rest of the community, and the innovator then becomes an informal resource for his/her colleagues.

Growers will require more ecological and economic information to manage agricultural and forest production systems. Because farms vary in their microenvironment, both spatially and temporally, growers that previously employed similar pest-management strategies may now use different systems. For instance, information of interactions among organisms within the agroecosystem can vary dramatically from the previous season's production system. Such information is more useful in a system which cuts across many interfaces, such as agroecology and economics, hence, enabling a grower to consider a wide variety of circumstances, drought versus a wet season, presence of specific pests, varying commodity and input prices, and a change in crop rotation. Part of an ecologically based pest-management system's success will be on an accompanying information system to help a grower manage the system.

Future pest managers may rely on computer based tools for greater efficiency and precision in decision making. More information on production system management is not only needed, it needs to be very accessible. Growers cannot lose time waiting for information on a timely decision nor can they lose time trying to find an information source. Though this information is not currently available for most agroecosystems, personal computers can improve the success of EBPM.

Researchers can provide growers with operational models to help growers make production decisions more directly. Some of the operational models now used, or that are being developed, are programs to use on a personal computer. An operational model is more effective when researchers work together to ensure accuracy of measurements used in the model; nonetheless operational models may be useful tools for decision making in EBPM.

Demonstrations are an effective means of reducing grower uncertainty to alternative management strategies. An effective demonstration of a new technology is probably the most powerful form of education. Pest-management demonstrations are conducted on university experiment station property or on private farms, often during planned workshops or field days. Such demonstrations also can be conducted by grower networks or cooperatives, Cooperative Extension personnel, or private business or other public or private agencies or organizations.

Demonstration research needs to be large in scale to reflect actual farming practices and to reduce outside influences (such as movement of arthropod pests and their natural enemies or plant disease spores). Whole-farm demonstrations are ideal, and these should be replicated throughout the area. Further, because local environment, pest pressures, and cropping practices vary from region to region, such demonstrations must be conducted in locations representing the entire range of the crop.

The agricultural information and education infrastructure has changed and adapted to new circumstances since the inception of the land grant colleges in 1862 (Goe and Kenney, 1988). The importance of the public sector in agricultural information was reinforced by the passage of the Smith-Lever Act of 1914, which created the Cooperative Extension Service, a unique cooperation of national, state, and local agencies for the primary purpose of providing practical information to growers. The current infrastructure will have to be retooled and strengthened to meet the knowledge needs for ecologically based pest management.

Historically, extension agents focused predominantly on recommendations for using pesticides, because of the large quantity of information available, the generally high degree of efficacy of the products, the ease of use of pesticides, rapid changes in pesticide use technology, and changes in pesticide use regulations. However, extension specialists recognize a need to increase their educational emphases in EBPM. A 1992 survey of 178 extension entomologists from throughout the United States, for example, found that 18 percent of their educational time was devoted specifically to programs relating to conventional biological control; the respondents indicated, however, that they expected the time would increase to 38 percent within 10 years (Mahr, 1995). This increase will be driven by research developments in biological control as well as by demand from the agricultural community for pesticide alternatives (Mahr, 1991).

Private businesses that rely on pest-management information can be placed in two categories: those with products to sell (such as agrichemical companies, biotechnology companies, and suppliers of biological-control agents), and those that provide a service (such as independent crop consultants). Both groups use information generated in-house as well from external private and public sources.

Product-oriented companies use scientific information (to further their own research on new products), regulatory and policy information, and marketing trends. Some of this information is also passed along to their customers, in the form of advertising, use recommendations, or product profiles. For example, the Association of Natural Biocontrol Producers, which is the association of companies which market predatory and parasitic insects for biological control, has developed a series of "Product Profiles" on the various types of commercially available natural enemies. Profiles provide information on the general biology of natural enemies and how they are best used in pest management. Information to develop the profiles derives from both public and private research.

Independent pest-management consultants sell their knowledge about pest control to growers and others who manage pests. As pest management becomes more biologically based, it is likely to become more knowledge intensive; it is therefore likely that private consultants will play an increasing role in pest-management decision making. Consultants require immediate access to new pest-management information, including new techniques and new products. In addition, they conduct their own research and use their own observations to evaluate how new practices and new products perform for their clients. Therefore, consultants routinely take advantage of all possible educational opportunities, including meetings, field days, publications, and trade journals; many of these opportunities such as extension programs, originate from the public sector.

The public sector, such as cooperative extension programs, must continue to have a major role in the delivery of pest-management information directly to growers and pest managers. The public sector is more likely to do environmentally oriented education than the private sector, especially with those important pest-management practices that are not related to commercial products or services. Also, the public sector, because it is not product or profit oriented, is perceived by the agricultural community as being less biased than the commercial sector when evaluating and teaching crop protection and production practices.

Pest-management methods that are effective, economically practical, long lasting, and not damaging to human health and the environment will require knowledge for effective implementation. As pest management becomes more biologically based, it is likely to become more knowledge intensive; this increased complexity will likely generate an increased demand for private consultants who can sell their knowledge about pest control to growers and others who manage pests. The numbers of these consultants is likely to be important for future information dissemination activities.

There is a need to assure educational and training opportunities for such consultants. The need for pest-management professionals has been emphasized by Mumford (1982). Edwards and Ford (1992) noted a lack of trained personnel to collect and process pest-management data for decision making—i.e., pest-management consultants. Alms (1994) advocated professional consultants with technical problem-solving skills who also can interact with the growers to help them change. The lack of well-trained resource people to work one-on-one with growers, either in the public or private sector, will constrain the rate at which ecologically based management systems are implemented.

Few universities have a fully integrated curriculum in plant health or pest management at either the undergraduate or graduate level. A few universities have developed such integrated programs in the past, especially at the M.Sc. level. Three reasons may partially explain the lack of success of such programs:

(1) they were too theoretical without building adequate practical experience into the program, (2) unfunded graduate programs could not compete with research programs providing stipends or assistantships to students, and (3) these programs may simply have been ahead of their time, coming during a period of relatively simplistic approaches to pest management that did not require practical knowledge needed by future practitioners of EBPM.

There is a need for greater coordination and better feedback mechanisms to ensure adoption of new EBPM strategies (Fitzner, 1993; Hoy, 1989; Norton, 1982; Zalom and Fry, 1992). All parties need to be involved in a coordinated process, including researchers, state and local extension personnel, the pest-management industry, end users, environmental groups, and administrators and policy makers.

ROLE OF COLLECTIVE ACTION IN PEST MANAGEMENT

Collective action can increase the successful development and implementation of EBPM. The previous section emphasized the need for information to decrease grower risk and increase adoption of these alternative approaches to pest management. Collective action strategies that unite growers or are the result of public policy can reduce costs and ensure the availability of effective biological organisms, products, or resistant plants.

Grower Cooperatives

Economically feasible solutions to solve pest problems often require the coordination and cooperation of growers of the same crop located within the same region. For example, the Fillmore Citrus District of Ventura County, California manages their supply of biological-control organisms and keeps pest-management costs below those of other districts (DeBach and Rosen, 1991). Growers are willing to pay for ecologically based management if they are assured that other members of the cooperative are paying their fair share too (Rook and Carlson, 1985).

Rook and Carlson (1985) assessed factors affecting the participation by North Carolina cotton growers in a private pest-management group over a three year period. The number of hectares in a time-competing crop, farm size, low costs of participation, and high expected cotton yield all contributed to the grower's decision to join the group. Growers that grew a time-competing crop found group management a way to free up time for the other profit making production system. Larger areas planted to the crop means greater vulnerability to damage and, hence, greater probability of joining the group. A cooperative can offer lower prices to growers which relieves some of the economic burdens associated with individual pest control and high cotton yields place a higher value on pest control.

The disadvantage of pest-management groups is that they are not necessarily set up to meet the unique demands of each grower. The quantity of pest control may be more or less than the needs of an individual grower. Thus, the more similar the needs and characteristics are among a pest-management group, the more likely that growers will join and participate.

Small-Market Support

Some of the most effective biological-control organisms and products in EBPM are projected to have modest uses or small markets, and will require public sector support for regulatory approval as well as for research. The best cases involve release of self-perpetuating classical biological-control organisms that produce great public good, but which have no commercial viability in the private sector. The costs associated with obtaining registration can discourage commercialization of biological-control organisms, as is the case now for minor use chemical pesticides, products needed on small acreage or in such small markets that they lack commercial appeal. In the same way that public funds help with approval processes of minor use chemical pesticides by regulatory agencies, public investment can ensure continued progress for biological technologies.

Certification

Organized efforts to increase distribution of high-quality biological tools will facilitate grower acceptance of EBPM. The procedure of crop seed certification by state agencies guarantees that new cultivars are genetically pure and that noxious weeds, arthropods, and pathogens are below detection limits. Each certification board tests, increases, releases, and distributes new field crops according to predetermined standards (Poehlman and Sleper, 1995). Such a process of release benefits the individual grower with valuable information on product performance and quality, lowering the risk of adoption.

Monitoring Pests

Monitoring for the potential development of resistance by pest organisms is key to managing the long-term viability of ecologically based systems (Gould, 1991). Biological-control organisms, products, and resistant plants are valuable entities and the numbers of these tools that meet the criteria of EBPM—safe, profitable, and durable—must be considered finite. Pest resistance to broad-spectrum, chemical pesticides is a recurring problem. Resistance to Bt, a microbial pesticide derived from *Bacillus thuringiensis* has also been observed in certain arthropod pests (Gould, 1991; National Audubon Society, 1991). The predominance of resistance biotypes will be directly related to the degree and duration of selection pressure applied to the target pest by the biological-control organism, product, or plant. If resistant biotypes selected from pest populations

Fillmore Citrus Protection District in California

The 3,500-hectare Fillmore Citrus Protective District in Ventura County, California, is a unique cooperative pest-management association. The district maintains its own insectaries to produce biological-control organisms as needed and uses only a minimum of selective insecticides. They have nearly perfected ecologically based management; as a result their pest-management costs are the lowest of any district in California and their fruit quality and quantity among the highest. Between 1971 and 1980, the mean annual cost of pest management in the Fillmore District was only $72 per hectare. This compares to $362 per hectare of Valencia oranges in Ventura County (Graebner et al., 1984).

The Fillmore Citrus Protective District originally was organized in 1922 to assist in the chemical eradication of the California red scale (*Aonidiella aurantii*), which had just been discovered in several orchards in the district. Prior to about 1940, most of the citrus orchards in the district were fumigated annually with hydrogen cyanide gas or sprayed with oil to reduce the Mediterranean black scale (*Saissetia oleae*), which was the major citrus pest in southern California at that time. Other pesticides were applied to control a variety of less critical pests.

The importation of a parasitoid, *Metaphycus helvolus*, from South Africa in 1937 to control black scale marked the beginning of a gradual shift to ecologically based management. This parasitoid has become established and provides good management of black scale. The original release of the parasitoid has been supplemented by periodic release of insectary-reared parasitoids in some Fillmore District groves. For the 10-year period of 1960 to 1970, an average of less than 5 percent of the groves were sprayed each year to manage black scale. This alone amounts to annual savings of $100 per hectare, more than $300,000 per year for the district.

The red scale parasitoid, *Aphytis melinus*, was imported from India and Pakistan and became widely established in the Fillmore District in 1961. Supplementary releases of insectary-reared *Aphytis* have been made as needed; nearly 4.5 billion parasitoids have been released since 1961. Since the parasitoid has been used, only about 1 percent of the district has had to be treated with insecticides annually for the California red scale.

In total, 12 arthropods that are serious major pests in other districts or other countries are being completely suppressed by use of biological-control organisms in the Fillmore District. Four of these were major pests at one time or another in the Fillmore District but now are either minor or innocuous due to importation of new biological-control organisms.

SOURCE: from DeBach, P., and D. Rosen. 1991. Biological Control by Natural Enemies. Cambridge, U.K.: Cambridge University Press, Second Edition. Pp. 372-373. Reprinted with the permission of Cambridge University Press.

Campbell Soup Company—
IPM Success on a Large Corporate Scale

Because of the greater complexity and more intense management needs of EBPM, questions have been raised about its probable degree of success in large-scale corporate agriculture. This example of the successful implementation of traditional IPM approaches by the Campbell Soup Company addresses this issue.

Background

Campbell Soup Company is a major producer and user of vegetables and other agricultural products. Its primary crop is tomato, but it also has its own mushroom and poultry operations. It contracts with vegetable growers in California, Ohio, Michigan, Texas, New Jersey, and Florida; much of its tomato production is in Mexico. Campbell recognizes the public's concern about pesticide use and residues in food. Therefore, in 1989 it embarked on an ambitious goal to reduce synthetic pesticide applications by 50 percent on crops grown for the company by 1994. Through its contracts, Campbell has encouraged its suppliers to use the latest IPM techniques. This program has resulted in substantial reductions in the use of pesticide and adoption of ecologically based practices.

Campbell's Approach to IPM

Campbell's IPM program includes three interrelated components. First, *cultural practices* include field selection and crop rotation to minimize weeds and diseases; vegetable varieties are selected based on their resistance to diseases; and field sanitation limits infestations of arthropods, weeds, and pathogens. Second, *monitoring* for pest activity is conducted at least weekly. The third component is *treatment*, which is implemented only when necessary and where appropriate, uses ecologically based practices such as sprays of *Bacillus thuringiensis* for caterpillar control.

Results

Anthrachnose Fruit Rot of Tomato in Ohio

Anthrachnose was the primary disease affecting tomato production in Ohio. Campbell adopted a computerized disease forecasting system, developed by the Ontario Ministry of Agriculture, that determines disease severity based on local weather conditions. Growers using this system used an average of 4.3 fungicide

become dominant, the usefulness of the biological control will be limited. The importance of monitoring pest populations in individual fields is critical to the accurate assessment of tactics aimed at delaying pest resistance to biological control.

Ensuring the durability of biological-control organisms, products, or resistant plants in the agroecosystem must be managed through collective action. An

Campbell Soup Company—continued

applications in 1991, compared to 9 applications for non-IPM growers, at a cost savings of $39 per acre. The quality of tomatoes did not decline in the IPM fields.

Virus Abatement in Mexican Tomato Production

In the 1980s gemini viruses were causing serious fruit losses in the state of Sinaloa, Mexico. These viruses are transmitted by whiteflies. Campbell developed a cooperative research program with California, Arizona, and Sinaloa. Geographic Information Systems (GIS) technology allowed for the identification of high-risk fields, which were then planted after whitefly populations dropped to low levels. More attention was paid to field sanitation, which eliminated alternate hosts for the whitefly. Using this approach, both pesticide use and disease incidence dropped substantially.

Lepidopterous Pests of Tomato in Sinaloa

A complex of caterpillars including tomato pinworm, tomato fruit worm, beet armyworm, and yellow-striped armyworm are prominent pests of tomato in Sinaloa. Traditional control of this complex required the use of up to 40 applications per crop of broad-spectrum insecticides. This led to detectable levels of insecticide residues. A variety of ecologically based practices have now been implemented, and their use is based on strict pest-monitoring practices. For example, pheromone mating disruption and microbial insecticides are used for tomato pinworm, and parasitic wasps are produced on site by Campbell and provided to growers for control of tomato fruit worm. This program resulted in the reduction in use of synthetic chemical insecticides in Sinaloa from 22,000 pounds in 1986–1987 to 0 pounds by 1992–1993, at a savings to growers of approximately $76 per acre.

In summary, through the adoption of effective monitoring techniques and the use of various EBPM practices, ranging from field sanitation to biological control, Campbell Soup Company has been able to increase yields, maintain quality, reduce pesticide use and detectable residues, and save its growers more than $1 million in pest-management inputs.

SOURCE: Adams, C. E. 1994. The role of IPM in a safe, healthy, plentiful food supply. Pp. 25-34 in Proceedings, Second National Integrated Pest Management Symposium/Workshop, Las Vegas, April 19-22, 1994.

effective mechanism currently used to delay pest resistance is the coordination of pest control activities among growers (National Audubon Society, 1994); sharing pest monitoring activities and collected information is useful in developing countermeasures that will limit additional crop losses. Practices, comparable to the use of mixtures and multilines to conserve the durability of plant resistance genes, can be developed and put into practice in order to enhance the durability of valuable biological-control organisms and products. Efforts to do this can be coordinated by industry, perhaps modelled on such organizations as the fungicide resistance working group, which is dedicated towards developing practices and implementation recommendations to enhance the long-term use of fungicides. Cooperative efforts among researchers, industrial suppliers, and growers can increase the durability of these biological tools.

Accelerating Research and Development

The natural interactions of an ecosystem supplemented with biological-control organisms, products, narrow-spectrum synthetic pesticides, and resistant cultivars form the basis of the EBPM systems proposed in this report. Implementation of these new pest control systems will require more knowledge about ecosystems than does the use of conventional pest management systems. Because the knowledge base in this area is limited, considerably more research will be required to develop and implement EBPM. Acquiring that knowledge base will necessitate coordinated efforts among many scientific disciplines.

The development of organic pesticides was a direct consequence of the enormous investment in organic chemistry research by national governments and industry since the late 1800s, beginning with dye chemistry research by German industry. Unfortunately, there is currently no such comprehensive investment in the kinds of research needed to help direct development of the various components of EBPM. EBPM research will benefit from the biotechnology knowledge base generated primarily through health science research; but a comparable knowledge base in ecology, particularly microbial ecology, is needed.

A full understanding of the dynamics among pests and control organisms in agricultural and forest ecosystems will require development of in situ methods to measure populations of both pests and control organisms and characterize their interactions—from multitrophic effects to molecular signaling. Managed ecosystems are complex and vary spatially and temporally. Pest and control organisms evolve and migrate; populations vary from field to field and among plants in an individual field. Diagnostic tools to measure pest populations must be useful both to individual growers and to researchers investigating pest management effects on a larger geographical scale.

Investigations will provide an understanding of the basis for the stability of these communities in natural systems as well as identify where the use of supplemental inputs and cultural practices disturbs the managed ecosystem and how pest populations develop and adapt to these disturbances. By more fully developing the ecological knowledge base and then coupling that with the expanding base of pest management experience, scientists can devise strategies to effectively manage pests and restore balance to forest and agricultural ecosystems. For example, research investigations of chemical signaling between insects and plants and production of toxic proteins by bacteria have been instrumental in developing biological-control products such as semiochemicals and biological insecticides.

Research is also needed to ensure that EBPM strategies can be transferred from the laboratory to a grower's field where agronomic practices and biological-control inputs can be evaluated in whole-farm settings. Moving from discovery to implementation requires a systems approach in which researchers from many disciplines cooperate in building safe, profitable, and durable pest management approaches.

FOUNDATIONS OF A KNOWLEDGE BASE

Currently EBPM research is the primary focus of a relatively small group of scientists whose contributions have led to the development of successful and economically feasible pest management systems. The challenge now is to move beyond optimal examples and into the mainstream of pest management. Accelerating the development of EBPM requires a clear agenda and institutions that can carry out that agenda. The focus should be to identify research that either overcomes obstacles to wider use of ecologically based management systems or leads to new and innovative ecologically based approaches.

Thus far only cursory knowledge about biological factors that control agricultural pest populations has been developed. For example, it is known that release of biological-control organisms to control pests is an option; however, the ability to predict the outcome of that release is limited. In fact, the proportion of releases of classical biological-control organisms that have resulted in complete suppression of the pest is rather low (Hall and Ehler, 1979; Hall et al., 1980).

EBPM implementation requires a basic understanding of the ecosystems in which agricultural predator and prey coexist, including an understanding of

• the many interacting factors that influence population size and the activities of control organisms and antagonists,

• the composition and dynamics of microbial communities present in soil and on plant surfaces,

• the food webs governing the population of pests, and

• the factors influencing the spatial distribution and differential fitness between weed and crop.

A knowledge-based approach to pest management requires an understanding of these factors, communities, and trophic levels—i.e., the interactions among organisms within the managed ecosystem. To predict an outcome of these interactions requires knowledge of

- vulnerable stages in organismal life cycles,
- factors influencing both pest and biological-control organism reproduction,
- disease vectors, especially vectors of viruses, and
- molecular signals governing pest and biological-control organism interactions.

It may not always be possible to find "natural" tools and tactics to control pests, neither will "natural" approaches always be the most ecologically sound. Many natural products are toxic to a broad spectrum of beneficial organisms and to humans; such products include neem, a pesticide extracted from the neem tree, and pyrethrum, an insecticide found in dried flowers of several Old World chrysanthemums. These natural compounds need replacement as much as do broad-spectrum synthetic chemicals. Because durability is an important goal of EBPM, organisms and chemicals that are specifically targeted to affect pest species but relatively benign to all other organisms must become the dominant management tools used in agricultural and forest systems of the future.

PRIORITY RESEARCH AREAS

A national research agenda, both general and cross-cutting, should identify broad areas of ecological research that promise to yield the critical information needed to accelerate progress in EBPM. The eight broad categories of research priorities identified by the committee can be used as a guide to the process of understanding how this complex, evolving system can be managed to reduce pests. These areas should receive priority for funding:

- research on the ecology of managed ecosystems;
- research on behavioral, physiological, and molecular mechanisms to effect EBPM;
- research to identify and conserve natural resources necessary for EBPM;
- development of better research and diagnostic techniques;
- development of ecologically based crop protection strategies;
- research on implementation and evaluation of EBPM;
- research to improve understanding of the socioeconomic issues affecting adoption of EBPM; and
- development of new institutional approaches to encourage the necessary interdisciplinary cooperation.

Research on the Ecology of Managed Ecosystems

Knowledge of managed ecosystems must guide the development of pest management tools. In the past 20 years progress has been made in answering questions about the ecology of managed systems; however, answers to old questions have generated new questions and challenges. The committee suggests that research in agricultural ecology over the next 10 years should focus on the areas noted below.

The Stability of Communities Containing Pests and the Communities' Sensitivity to Perturbation

Probably the most important research needed to learn how to manage pests is research to understand the basis for stability of plant, arthropod, and microbial communities. Soil-inhabiting pests and plant-parasitic nematodes present special problems because most species of soil microbial communities have not yet been catalogued. Because many species that form ecosystem communities are not known, community dynamics and effects on soil pathogens and nematodes are not well understood. In many cases, it is not known whether ecological theory developed for macroorganisms also applies to microbes. A recent treatise on the subject declares that "microbial ecology is experimentation in search of theory" (Andrews, 1991).

To the extent they can be characterized, microbial communities appear to be altered only transiently in response to the introduction of biological-control organisms. The basis of this stability in community structure is not likely to be understood until the major components of the community have been identified and their roles elucidated. With the identification of the components of soil communities, and their functions, new organisms may be identified that, on augmentation, could serve as effective biological-control organisms.

Most studies of nonpest soil invertebrates have focused on determining how the diversity and abundance of species is affected by specific changes in cropping systems (e.g., tillage and pesticide use), but very little is known about how the sometimes significant changes in the species composition of these communities affects survival and growth of pest populations. Studying organisms in the soil milieu has been extremely difficult using conventional methodologies, highlighting the need to develop new tools for these investigations.

Physical, Chemical, and Biological Conditions in the Rhizosphere and the Phyllosphere and Their Effect on the Development of Pest Populations

Soil conditions affected by cultural practices and chemical inputs (e.g., fertilizers and pesticides) in turn affect the activities of microorganisms, arthropods, and weeds. Unfortunately, it is not known which specific soil characteristics have the most significant impact on these organisms. For example, it is known that green manures enhance general microbial activity in soils planted to potatoes and result in the suppression of Verticillium wilt (Davis et al., 1994; Schroth et al., 1992); however, the specific mechanisms by which this process occurs are poorly understood.

More is known about the components of microbial communities in the phyllosphere than in the rhizosphere, yet there is still much to learn about the conditions that affect the microbial populations in the phyllosphere. On leaf surfaces, the population density of bacterial foliar pathogens is known to undergo striking changes in response to the environment (Hirano and Upper, 1990), but little is known about how population dynamics of such pathogens affect other components of the phyllosphere community or vice versa. Competition is thought to play an important role in determining the dynamics of microbial populations in both the rhizosphere and phyllosphere (Loper and Lindow, 1993), but science has only hints as to the nature of this competition.

Attempts have been made to reduce frost damage caused by bacteria that nucleate ice on plants by introducing strains of that same bacterium that lack the gene responsible for the ice-nucleating activity. Such competition experiments with bacteria, engineered using recombinant DNA technology to remove the ice-nucleating genes, were the first reported experiences in which recombinant organisms were released in the environment (Lindow, 1993). These experiments focused public concern on issues related to the use of genetically engineered organisms to solve agricultural problems. Although these experiments served to draw attention to the issue of risk associated with biological control, they also showed the potential of this new technology to build an increased understanding of microbial ecology.

The Role of Dispersal Mechanisms in Establishment of Populations of Major Pests and Their Biological-Control Organisms

Many agriculturally important weeds, invertebrates, and microbes are not permanent residents of the crop fields and orchards they inhabit. Because of annual (or longer term) changes in the suitability of specific fields or geographic regions for survival of these species, many persist based on their ability to physically shift the location of their populations. The movement involved in these shifts can be limited to a few kilometers or range from the southern to northern boundaries of the United States. Some of this movement involves establishment of only a small number of individuals to initiate populations in new locations. In other cases, however, millions of microbial spores or swarms of arthropod pests can invade an area within a few days and cause rapid destruction of a crop.

Although progress has been made in understanding the movement of many key pests, information is still needed to accurately predict the timing and direction of that movement. Such predictive power can provide farmers with advanced warning of invasions, which in turn may persuade many growers to use responsive instead of prescriptive pest management methods. Furthermore, a more complete understanding of movement of beneficial organisms can lead scientists to develop methods to promote efficient movement of these organisms between heterogeneous agricultural habitats and between agricultural and non-agricultural areas.

Inoculation of Agricultural Systems with Specific Biological-Control Agents: Effects on Nontarget Organisms

There has been much concern about the nontarget effects of conventional pesticides; but as biological organisms, products, and resistant cultivars become the centerpiece of pest management, more attention will need to be focused on their potential side effects. When an introduced biological-control organism fails, it will be necessary to determine why it failed, rather than simply search for alternative agents. It will also become extremely important to know whether an organism slated for introduction has the potential for long-lasting negative effects.

There has been justifiable concern about introduction and release of non-indigenous and genetically engineered biological-control agents. There are many unanswered questions about the ecological effects of these organisms; and in most cases the techniques or tools to properly obtain this knowledge do not exist. Thus it will be necessary for ecologists and molecular biologists to conduct interactive interdisciplinary research to develop new tools for studying the biology of beneficial organisms. In situ hybridization and new methods for analysis of DNA extracted directly from environmental samples are expected to provide further insight into the distributions and populations of beneficial microbes.

Manipulating the Spatial and Temporal Structure of Agricultural Habitats to Reduce Pest Outbreaks

An agricultural habitat's suitability for survival of invertebrates and microbes varies greatly over both time and space. Change over time is especially pronounced in temperate zones where agricultural crops are rotated on a seasonal basis; the change in mean temperature combined with the change in crop species effects a dramatic, wide-scale habitat alteration for most organisms. Agricultural habitats also can vary in suitability on the scale of meters and centimeters. Plant pathologists and entomologists have long known that outbreaks of some pests are directly related to the spatial layout of crop fields as well as the location of fallow land. Weed scientists who have studied the spread of nonindigenous weed species know that land use patterns influence invasion and establishment of these species. Landscape ecology, which took shape as a formal subdiscipline of ecology in the early 1970s, focuses specifically on the response of individual plant and animal species and communities to distinctive landscape patterns. Geographic information systems (GIS) and remote sensing technologies—new tools in this area of study—are leading to rapid advances in understanding how temporal and spatial variation affect survival of endangered species, pest populations, and beneficial organisms.

One technique of landscape ecology used by weed scientists involves manipulating spatial access to water and nutrients in order to suppress weed populations and increase crop yield. Thus, techniques and tools developed by landscape ecology scientists and traditional techniques in life-system ecology can be used to identify subtle changes in some agricultural landscapes that could improve the competitive advantage of beneficial organisms while negatively impacting pests. The effects of specific changes in the agricultural landscape on the overall durability of any local cropping system can only be adequately addressed by interdisciplinary teams including weed scientists, plant pathologists, entomologists, biochemists, economists, sociologists, crop scientists, etc.

Biological Characteristics that Enable Organisms to Adapt to Changes in Physical, Chemical, and Biotic Conditions

It is known that repeated use of pesticides conditions many targeted pests to naturally select for resistant biotypes. Currently, the arthropod class contains the largest number of species to have developed resistance to pesticides. However, weeds developing resistance to herbicides is a growing problem in crop production; in fact, researchers have discovered plant pathogens that can overcome resistant cultivars within just a few growing seasons. Although the mechanisms and dynamics of pesticide resistance have received increased attention over the past 40 years, more research is needed to understand how both detrimental and beneficial agricultural organisms adapt to their environment. The little information that does exist on this topic indicates that organisms in agricultural and forest systems readily adapt to the many abiotic and biotic changes in their habitats.

Classical tools of ecological genetics and population genetics, and new tools of molecular biology, are being used by researchers in attempts to slow the rate at which pests adapt to chemical and physical stresses and, at the same time, increase the rate at which beneficial predators adapt to these stresses and to the defenses of their target prey. For example, Lewis and Martin (1990) found that preconditioned parasitoids are more effective than nonconditioned parasitoids when released into the environment.

Better Ways to Understand How Multispecies and Multitrophic-Level Interactions Contribute to Pest Population Dynamics

Plant pests are components of ecological systems. Their presence is made possible by many ecological factors, and their effects result in multiple ecological interactions. Concerns related to multispecies and multitrophic interactions are leading to some experimentation and field applications.

  • Plant pathologists are infecting plants with certain minor pathogens that protect the plants from more devastating pathogens.
  • Weed scientists are investigating how the interactions of a crop infested with multiple species of weeds affect yield.
  • Research teams are discovering higher concentrations of beneficial organism attractants in wild plants than in their commercial counterparts.

The knowledge base in this area has expanded in the past decade, but a complete understanding of such multitrophic-level interactions is essential for designing effective and reliable biological-control measures for EBPM.

Research on Behavioral, Physiological, and Molecular Mechanisms to Effect EBPM

The mechanisms that govern interactions among organisms have been an area of intense study in both medical and agricultural research because understanding these mechanisms is key to understanding how pests and diseases can be controlled through active intervention. For agriculture, one of the pay-offs of such knowledge would be the ability to quickly develop genetically engineered plants resistant to pests. Already the toxins responsible for the success of *Bacillus thuringiensis* (Bt) as a biological-control agent offer promise for the enhancement of plant resistance to specific arthropods when the genes encoding the toxins are incorporated into plant genomes.

Knowing what factors control interactions among organisms will help scientists better understand how to manipulate pest populations, or communities containing these organisms, to reduce their impact. Understanding biological control at the cellular or molecular level has a predictive value, enabling scientists to set realistic expectations for when and where biological control may succeed or fail. For example, if biological control is known to be mediated by iron competition between an antagonist and a target pathogen, it is not likely to succeed in conditions in which iron is replete. Identification of the characteristics that determine biological-control activity is prerequisite to manipulating these characteristics through genetics, modification of behavior or the environment, or other interventions.

The discussion of research needs in the following section provides suggestions as to how additional knowledge of processes directly related to biological control can be used to enhance biological-control processes. The committee's recommendations are meant to focus on gaps in current research priorities that are uniquely related to biological-control issues. Although there have been some major developments in studies of plant-host interactions (e.g., the finding that a specific plant host, by cellular signaling, indicates recognition of a pathogen), such issues as arthropod signaling, antagonism/toxicity/ antibiosis, host selection, target-host diseases, target-host parasitism, and plant-disease resistance are subjects in need of understanding at behavioral, molecular, and physiological levels.

Manipulating the Behavior of Pests to Reduce the Risk of Pest Outbreaks

Modifying pest behavior to reduce pest damage is a more long-lasting approach to pest management than pest destruction; however, success with behavior modification of pests and natural control agents has been restricted to a limited number of cases. There have, however, been some advances made in molecular biology and behavioral studies that have highlighted the need for researchers to cooperate on signaling system projects. Tumlinson and his colleagues formed an interdisciplinary team to investigate plant, herbivore, and parasitoid interactions (Tumlinson et al., 1993). They found that oral secretions of herbivores activate plants to emit a set of volatile signals that benefit both the plant and parasitoids by alerting the parasitoids to the presence and location of their herbivore hosts. They further report that these biological-control organisms learn to process visual and olfactory information to more efficiently locate plant pests.

Although "behavior" is generally considered to be a property of multicellular organisms, microbes also respond to signals that control behaviors such as mating, "feeding," movement, and germination; so behavior-based management techniques could also be developed for some microbes.

Knowledge that some arthropod and nematode pests must locate and recognize a crop as a food source can guide scientists to chemicals capable of controlling pest foraging behaviors. An understanding of organisms' behavior in managed ecosystems can also help scientists design cropping systems that favor the success of biological-control organisms and reduce pest damage (Kareiva, 1990; Lewis et al., 1990). For example, arthropod pests can be lured away from a valuable crop by planting a less vulnerable or less valuable trap crop nearby. Knowledge gained from manipulating the behavior of arthropod pests could be useful in research to manage weeds and pathogens.

Signaling between Target Hosts and Biological-Control Organisms

There is much interest in identifying signals and elucidating the nature of signal reception and transduction in organisms. Chemical signals influence an organism's growth and development and affect interactions between different organisms, such as in the induction of host-defense responses. Much is known about the role of signal molecules and physical cues in determining the behavior of arthropods; but more comprehensive knowledge of the nature and function of these signals is needed if they are to be exploited for biological control of pests, and investigations should address the behavior of the pests in situ, not just in the laboratory.

One role of signaling between different organisms within communities can be seen in the response of pests to signals from potential host plants. For example, acetosyringone released from wounds of certain plant species induces the transcription of the *vir* genes of Ti plasmid-containing strains of *Agrobacterium tumefaciens*, initiating the process of crown gall formation (Gelvin, 1992). It is reasonable to assume that similar signaling occurs between many of the co-evolved components of rhizosphere and phylloplane communities.

One of the most useful applications of chemical signaling has been the use of pheromones for trapping and quantifying populations of specific arthropods. Although there has as yet been no comparable practical applications in the control of plant pathogens, pheromones are known to control fungal sexual cycles (Spellig et al., 1994). With better understanding, fungal pheromones can be used to control fungal behavior.

Using Natural Antagonism, Toxicity, and Antibiosis to Control Pests

During the past decade, compelling evidence of the ecological roles of natural antibiotics produced in situ by microorganisms has been reported (Weller and Thomashow, 1993). Whereas many scientists formerly believed that antibiotic compounds were produced as artifacts of laboratory culture conditions, it is now known that such compounds can be and are produced by microorganisms—e.g., *B. thuringiensis* and *A. radiobacter*—inhabiting natural substrates and in concentrations adequate to inhibit pests. Microorganisms that produce antibiotic compounds are among the most successful biological-control agents. Future research could lead to

* production of novel antibiotics through domain switching, as has been done for polyketide antibiotics;
* understanding when genes encoding antibiotic biosynthetic enzymes are expressed in situ and what environmental cues induce antibiotic biosynthesis;
* optimizing in situ antibiotic production by placing biosynthetic operons under regulatory control of constitutive promoters or by modifying the biological environment to enhance biosynthetic gene expression; and
* evaluation of pest resistance to antibiotics or toxins produced by biological-control agents.

To date, manipulation of genes encoding the δ-endotoxin (Bt toxin) of *B. thuringiensis* has received the most attention for practical application as a biological insecticide. *B. thuringiensis* produces massive amounts of these crystalline insecticidal proteins, which differ in shape, size, and host-specificity. Bt-toxin genes are being introduced into plants as resistance genes; into endophytic bacteria that reside in the xylem of plants; and into bacteria that, after heat-killing, serve to encapsulate the toxin, thereby enhancing its persistence in the field. Some products using these strategies are currently commercially available, others are in the development stage. The success of Bt-toxin research raises optimism that other organisms are also potential sources of effective biological-control products.

Using Pathogens of Plant Pests for Biological Control

Pathogens of diseases that afflict pests offer potential for use as biological controls. Unfortunately, the time lag between when the disease is introduced and when it becomes established in the pest population limits this method; crop damage during the incubation period often is too extensive to be economically acceptable. This is the problem with baculoviruses, which otherwise offer considerable potential for biological control because of their specificity and safety. Research to identify the initial steps of baculovirus pathogenesis can lead to the development of fast-acting baculoviruses. Other research is being directed at developing disabled ("suicide") recombinant baculoviruses which would eliminate only the arthropod that received the virus. Disabled viruses are attractive because there is only a remote possibility that a foreign gene introduced into the viral genome by recombinant DNA technology could enter the ecosystem. Research plans are also being developed to determine whether some microbial pathogens that have coevolved with arthropods could be manipulated to control specific arthropod pest populations.

A virus of the plant pathogenic fungus *Cryphonectria parasitica* is an important model to describe how naturally occurring or recombinant viruses can control fungal diseases of plants (Nuss, 1992). Work with this biological-control agent has shown the potential for use of viruses to control fungal diseases. Although use of viruses of fungal pathogens could result in a reduced virulence of the fungal pathogen population, much more information about viruses is needed before their potential for biological control can be exploited fully.

Perhaps the greatest potential for the use of diseases as biological controls is in weed management. Research is needed to enhance or conserve naturally occurring soil microbes that have weed suppressive abilities. Unfortunately weed management via microbial agents will not be viable until it is possible to (a) increase the speed and effectiveness of the control agent in the field and (b) ensure that the agent will not damage beneficial plants. Understanding the genetic basis of pathogenicity, virulence, and host-range of microbes that attack weeds could result in the selection and development of safer and more effective microbial weed-control agents.

Improving Strains of the Parasites, Parasitoids, and Predators of Plant Pests

To date, the major emphasis in genetic improvement of arthropod predators

Biological Control of Chestnut Blight

Will blight end the chestnut?
The farmers rather guess not.
It keeps smoldering at the roots
And sending up new shoots
Till another parasite
Shall come to end the blight.

Robert Frost

The American chestnut tree once comprised more than 25 percent of the eastern hardwood forest. It persists today in its native range only as sprouts or as decaying remnants of the past. Soon after the discovery in 1904 of *Cryphonectria parasitica* on an imported Oriental chestnut in New York, the imminent demise of the American chestnut from chestnut blight was well recognized. In the 1930s, the pathogen was observed in Europe, where the European chestnut also succumbed to chestnut blight.

Approximately two decades after chestnut blight was first diagnosed in Europe, certain trees in northern Italy appeared to recover from the disease. Isolates of *C. parasitica* obtained from healed cankers were hypovirulent: they infected European chestnut but rarely produced lethal infections. Later, stands of American chestnut that survived blight were found in Michigan. Many isolates of *C. parasitica* obtained from healed cankers of Michigan trees also were hypovirulent. Hypovirulence is caused by a double-stranded RNA (dsRNA) virus that infects the fungus, reducing its virulence on chestnut trees.

When hyphae of a hypovirulent and virulent isolate of *C. parasitica* fuse, the virus and other cellular contents can be transferred from the hypovirulent to the virulent isolate. On acquisition of the virus, the virulent isolate becomes hypovirulent. In chestnut trees, transfer of the virus from the hypovirulent to the virulent form occurs by this process and, as a result of viral infection of the virulent fungus, the tree can actually recover from chestnut blight. Dissemination of the dsRNA virus among isolates of *C. parasitica* can occur naturally; natural dissemination restored the chestnut as a dominant forest species in northern Italy. Naturally recovering stands in Michigan are also testaments to the success of transmissible hypovirulence for control of chestnut blight.

Despite the presence of hypovirulent strains in the hardwood forests of the eastern United States, natural recovery of the chestnut forest has been slow and geographically limited. Fusion of hyphae, and subsequent transmission of the dsRNA virus, can occur only between isolates of *C. parasitica* within a single group (called an anastomosis group). The population of *C. parasitica* is very diverse in the eastern United States, and the presence of multiple anastomosis groups is one factor slowing the dissemination of hypovirulence to native virulent isolates in this region. Recently, a transgenic isolate of *C. parasitica* containing the viral genome was constructed. Transfer of the viral genome from the transgenic isolate to other isolates of *C. parasitica* does not depend exclusively on hyphal fusion. Therefore, the transgenic isolate may be an effective source of the virus for transmission to virulent isolates of *C. parasitica* in the eastern United States. Because *C. parasitica* does not infect the root systems but only the shoots of the chestnut tree, the genetic diversity of the eastern chestnut forest has been preserved throughout this century.

SOURCE: Nuss, D. L. 1992. Biological control of chestnut blight: An example of virus-mediated attenuation of fungal pathogenesis. Microbiol. Rev. 56:561-576.

and parasitoids has been on developing natural control agents that are resistant to pesticides so that they might function within a system dominated by broad-spectrum pesticides. In the future, both classical breeding techniques and genetic engineering could be used to increase the efficiency of biological-control agents in specific agricultural systems. Characteristics that could be altered include heat and cold tolerance, host range, and migratory cues.

Certain fungi trap and feed on plant-parasitic nematodes. These and fungi that parasitize other fungi, such as *Trichoderma* spp. and *Gliocladium* spp., are useful for biological control of soil-borne diseases in greenhouses (Lumsden et al., 1993). More can and should be done to determine the importance of parasitism in the biological-control activity of these fungi and to characterize the molecular basis of parasitism. As with other biological-control agents, more information is needed about their inherent genetic variability as well as information about which characteristics of the biological-control organism are important targets for strain improvement.

The Basis of Host Selection and Host-Range Specificity

An important concern of the general public about the release of non-indigenous predators and parasitoids is their potential to harm indigenous nonpest species. To allay these concerns requires a better understanding of the genetics of host-range specificity, which will provide insights into how organisms can be genetically altered to limit their current and potential host range and thus enhance their safety as control organisms.

Some progress has been made in understanding the genetic basis of host selection by microbial plant pathogens—an important step toward understanding the molecular basis of host specialization by plant pathogens. In fungal, bacterial, and viral plant pathogens, genes that determine the spectrum of plant cultivars that pathogens can infect have been identified. Staskawicz and colleagues (1984) changed the host specificity of a selected pathogen by deleting or inactivating a single gene. The host range of a plant pathogen was narrowed genetically by Mellano and Cooksey (1988).

The genetic basis for host selection and host specificity in arthropod predators and parasitoids is less understood than it is for plant pathogens, but progress is being made in this area. It is known, for instance, that host specificity of an arthropod parasitoid is determined by a virus that the parasitoid transmits to its host when it inserts one of its eggs into the host. It is also known that within a given host range, arthropod parasitoids can adjust their host preferences by learning cues that lead to a suitable host. Although predators with a broad host range have preferences for specific prey, the genetic factors that govern preference are not understood. Research in the area of host selection could have long-term benefits in improving the efficiency and safety of these biological-control organisms.

Pest-Resistance Mechanisms of Plants

Host plant resistance is the most important component of many plant disease management systems. The development of molecular biology research tools has resulted in identification and analysis of genes that control the interactions of plants and pathogens and is contributing to efforts to produce transgenic crop plants resistant to a variety of plant pests.

Although resistance is less central to management of arthropod pests than of diseases, it is a major component of certain control programs such as the wheat-hessian fly and small grains-greenbug systems. Such programs may become more important for all crops when research on genetic engineering of arthropod-resistant plants gains momentum. Research has focused on expressing toxic proteins in crop plants to kill pest species. Scientists are now limited to use of a few toxic proteins such as Bt toxins, but this may change based on (a) screening of novel microbes for active compounds and (b) research into physiology and biochemistry of metabolic and hormonal systems unique to certain taxa of arthropod pests. The recent discovery of an insecticidal cholesterol oxidase produced by a fungus offers the possibility that other such toxins will be discovered. The fact that the activity of Bt toxins and other protein-based toxins is limited to arthropods is encouraging. Industry is involved in discovering novel toxins, and academic research in arthropod physiology and biochemistry is likely to help identify arthropod-specific targets and novel approaches for engineering plants.

Plants can respond to infection by an active resistance response, termed *induced resistance*, which is maintained systemically throughout the plant for a period of time after the infection. This response can protect the plant from some subsequent infections by plant pathogens. Recent progress in understanding the mechanism of signaling the advent of such challenges between different parts of the plant offers hope that this phenomenon will soon be explained at the molecular level (Kessmann et al., 1994). An understanding of the process could lead to novel, biologically based means to control damage caused by some pests. Clearly, basic studies in plant biology can have important impacts in agriculture. Continued support of research in this area is required if the full potential of these and other approaches is to be realized (National Research Council, 1989b).

Research to Identify and Conserve Natural Resources Needed for EBPM

A founding principle of classical biological control of exotic pests is that natural enemies of the pest can be found in the geographic region where the pest evolved. Likewise, plant geneticists have found that the best place to find pest-resistant plant varieties is in the geographic region where the plant and pest coevolved. Collection and identification of germplasm from such regions has been a high priority for plant breeders. It is probable that every pest has at least one biological-control agent that could be identified by this approach.

Systematic exploration for biological-control agents, however, has been limited to a relatively small number of arthropod and weed pests. For fungi, bacteria, and nematodes, even less exploration has occurred. Research to identify, collect, and document the natural enemies of pests in their native habitat should continue to be an important component of efforts to discover useful biological-control agents.

Preserving biodiversity is justified by the benefits to be accrued from identifying beneficial products, such as taxol; and the economic benefits of preserving potential biological-control agents as part of biodiversity are expected to be significant. A classic example is the discovery of the pesticidal properties of the neem plant; its extract acts as an "antifeedant," eliminating the urge to eat for a variety of arthropod pests. Random screening for such chemicals would probably not be cost effective; a logical and more cost-effective alternative would be enlisting the aid of indigenous people knowledgeable about the regional flora to identify unique biotic products that have known pest-control properties (Thurston, 1990).

Systematics Research and Taxonomic Resources

Reliance on conventional chemical pesticides for suppression of agricultural pests has obscured the need for continued development of systematics or taxonomic resources; however, careful characterization of pest populations is important to the success of the more target-specific methods of EBPM. Taxonomists are continuously challenged to accurately identify an increasingly diverse group of pests and potential biological-control organisms that includes both arthropods and plant pathogens. Unfortunately, systematics research has not kept pace with the growing demand, and many of the most important groups of pests and biological-control agents are poorly characterized. Even in agricultural systems that have been studied for decades, many natural enemies of agricultural pests do not fit into described species and therefore their classification remains uncertain. In unmanaged ecosystems where effective natural control agents are found in association with their hosts, the number of undescribed species is even higher.

Revised classification at the family, genus, and species levels is needed for a broad array of important pest and antagonist groups. The present taxonomy of microorganisms is oriented to medical clinical isolates, a situation that presents a special problem for biological control as regards regulation. Because the taxa are structured to characterize clinical isolates, many isolates from nature that cause no known clinical problems are often labeled as pathogens because there is no related nonpathogenic taxon in which the isolate can be placed. The problem has been exacerbated by the widespread use of commercial data base systems that are almost entirely focused on identification of clinical bacteria. Accurate identification of organisms used by pest management researchers depends on an understanding of the systematics of the organism and on the availability of tools to make accurate nominative determinations. Unfortunately, revised classification at the family and genus levels may require 10 to 20 years to complete, and few such studies have reliable financial support.

Because populations of pests within a species are heterogenous, characterization of the organism is, in many cases, needed at the subspecies or "strain" level, where conventional alpha taxonomy has not provided sufficient support. Molecular, numerical, and other advanced biological tools are now available to address some of the heretofore intractable problems in systematics such as strain characterization and species discrimination. Tools including DNA amplification, DNA sequence analysis, hybridization studies, restriction fragment length polymorphisms, and various advanced forms of electrophoresis are increasingly used in research laboratories. Using these tools could provide the taxonomic foundation essential for development of EBPM.

Development of Better Research and Diagnostic Techniques

Researchers should devise new methods to study, monitor, and evaluate agricultural and forest system processes as well as the effectiveness of pest management tools. Providing researchers and farmers with improved diagnostic techniques is key to both developing and increasing the use of biological organisms, products, and resistant plants. The examples provided here are not comprehensive, but illustrate how development of methods and techniques are needed to understand and implement EBPM strategies.

Methods for in Situ Study of Microbial Populations

The lack of methods for studying the growth and development of microbes in situ hampers implementation of microbes as biological-control agents. Although it is recognized that only a small percentage of microbes present in natural habitats can be cultured (Oliver, 1993), current methods rely almost exclusively on culturing techniques to isolate, enumerate, and characterize microbial control agents and target plant pathogenic bacteria and fungi. Novel approaches based on in situ hybridization, immunoisolation using monoclonal antibodies, or amplification of DNA isolated directly from environmental samples offer promise for the in situ study and quantification of microbial populations. Future studies using such techniques will greatly enhance understanding of the composition of communities of plant-associated microbes and population sizes of introduced antagonists or weed or insect pathogens as well as to increase understanding of the complex interaction between soil microbes, allelopathy, and resulting weed-suppressive soils.

A further constraint to the implementation of microbes as control agents is the inability to evaluate their in situ activities, including the production of critical metabolites or expression of phenotypes essential for biological-control activity.

Many phenotypes that contribute to the biological-control activity of such microbes can be readily detected and quantified in culture but are difficult to assess in situ with current methods. The development of reporter gene systems (i.e., a gene encoding a phenotype that is readily detected and quantified in natural habitats serves as a "reporter" of the activity of a gene that is critical to biological-control activity) offers promise for the assessment of in situ activities of microbial control agents (Lindow, 1995). Creative development and application of new methods for the detection and characterization of microbes in situ must be encouraged.

A significant limitation to progress both in understanding biological-control processes and in developing better biological-control agents is the lack of effective methods to genetically manipulate these organisms. Understanding has advanced based on scientists' use of model systems. Unfortunately, the tools developed to genetically manipulate these model systems often do not work with organisms of interest for biological control. Transformation methods and vectors need to be developed for organisms potentially important for biological control.

Grower-Friendly Diagnostic and Monitoring Methods

EBPM will require a substantial input of information at the farm level, and its success will depend on accessibility of input procedures. The development of diagnostic kits that can be conveniently used and reliably interpreted for measuring and monitoring populations of pests and natural enemies, detecting weed seeds, and assessing pest contamination of planting material are examples of areas in which research is needed to develop usable procedures.

Production, Stabilization, and Delivery of Biological-Control Organisms and Products

An important challenge to implementation of EBPM will be to provide reliable sources of living biological-control organisms. Because most of this information is maintained as carefully guarded trade secrets, it is difficult to assess the status of efforts developed to maintain the viability of living biological-control organisms and products in commercial formulations. There is more of a history of private-public sector cooperative research in fermentation or rearing of biological-control agents, but it is still largely funded by industry. Opportunities for cooperative research exist also in developing methods to diminish production of phytotoxic or otherwise deleterious metabolites of microorganisms and in identifying optimal physiological conditions for the storage of biological-control agents. For example, it is thought that Gram-negative bacteria, which lack the resistant spore state of many Gram-positive bacteria and fungi, cannot withstand formulation and have a short shelf life. There has been, however, research indicating that the "stationary phase" may be a stress-resistant phase of the Gram-negative life cycle (Kolter et al., 1993) that is analogous in many respects to spores of Gram-positive bacteria. Future research may be directed on genetic selection of biological-control bacteria with traits for stress resistance.

Another obstacle to implementation of biological control is the large amount of inoculum required for efficacy of certain microbial agents. Research to optimize placement of biological-control agents in close proximity to target pests and production of propagules at the most effective physiological state can do much to remedy this problem. These are the types of issues that could be fruitfully explored through cooperative private-public sector research efforts.

Development of Ecologically Based Crop Protection Strategies

Previous research priorities have addressed components of various biological-control strategies, however, the goal of durable pest control can only be achieved if the various strategies are coordinated and implemented in an integrated manner. The concept of IPM provided a good model for the type of integration that would be desirable, but IPM also serves to exemplify pitfalls to be avoided. For EBPM to succeed, there must be strong input from scientists involved in all pest-related disciplines.

Because pest management strategies are linked to other farming system components, it will be necessary to study relationships among, for instance, pest management, plant nutrition, and farming practices. By doing so, researchers will gain comprehensive data—essential for the development of an integrated crop protection strategy. The focus of this section, however, is directed toward EBPM-related research.

Develop Predictive Models for Cropping Systems

Integrating biological and cultural (physical) methods of pest and disease control is key to achieving long-term pest management. The biological/natural concept for managing pests has historically been ecologically based. Such an approach recognizes that managed cropping systems exist within and are a part of an ecological web of relationships where physical and biological processes are interactive and dynamic. The plants, herbivores, predators, pathogens, weeds, etc., that exist within this web have developed a repertoire of offensive and defensive maneuvers and countermaneuvers in response to one another. At present, the primary limitation to durable cropping systems is insufficient knowledge of the relative importance of system interactions often in and of themselves as well as in relation to the system as a whole.

Science needs to be able to predict the consequences of the various potential interactions within a single cropping system. Individual interactions must be understood, and attempts to link the various biotic/biotic and biotic/abiotic processes into a predictive system must be made. Integration will be a challenge both because of the complexity of the problem and because of the requirement that different disciplines must coordinate efforts and work together. Models can provide a theoretical basis to predict the interactions of the various biotic and abiotic components of an ecosystem. However, the knowledge gained must be applicable to agricultural and forest production systems.

The type of research needed is much like that addressed in IPM studies. For EBPM, however, biological-control practices must be integrated with each other and with normal crop-management practices. Such integration will require a variety of investigations:

- timing of applications,
- methods of application,
- amount of biological control applied, and
- whether controls can be applied simultaneously and what would be the outcome.

For example, although there is well-justified interest in developing microbial mixtures for biological control, it has been observed that certain strains degrade the antibiotics produced by a second strain when the two strains are mixed together. In this case, the combination of the two strains may be less effective than if each strain were applied individually. An integrated scientific approach can direct research efforts toward practical pest-management solutions.

Develop Strategies for Achieving Durable Plant-Host Resistance

Concern has been expressed about the durability of pest-resistance genes, but pathogen- and arthropod-resistant cultivars can be deployed in ways that reduce or eliminate the breakdown of resistance caused by changes in the genetics of pest populations. Innovations in genetics and applied evolutionary biology provide an excellent opportunity for researchers to develop methods to augment the few commercial practices available for protection of pest-resistance genes. There is a need to increase scientific collaboration between plant molecular biologists, pest molecular biologists, evolutionary biologists, and crop scientists so that real progress can be made.

The durability of resistance genes has long been an area of active research that has led to strategies of deployment of multigenic resistance, multiline varieties, and varietal mixtures. The dangers inherent in the genetic vulnerability of crop plants will need to guide both research and commercial management of resistance genes. This will be particularly true of resistant genes introduced into transgenic plants because they may be widely distributed among plant taxa. With the discovery and use of these genes should come studies investigating ways to increase their stability after deployment.

It is also important that assessment of resistance genes not be limited to only direct effects on plant pests. Resistance genes may also affect natural enemies of the pests and microbial communities in the rhizosphere or phyllosphere. In the manipulation of plant genetics, thought should be given to how traits of crop plants can be altered to enhance the overall health by enhancing populations or activities of naturally occurring or introduced biological-control organisms.

Research on Implementation and Evaluation of EBPM

The research needs for EBPM discussed in the sections above have emphasized the need for new knowledge of biological processes, interactions, or organisms useful for biological control. Emphasis should be placed on implementation research to increase adoption of EBPM. There is a need for applied implementation research, as well as farm-scale and area-wide evaluation of the biological and socioeconomic impacts of new management tactics.

Implementation Research

Large-scale field trials are important to gaining grower acceptance of EBPM. In a survey of entomologists in the 12 north-central states, the lack of funding of farm-scale research was identified as one of the major constraints to the implementation of biological controls in pest management (Mahr, 1991). A large-scale trial can be performed on an operational farm where researchers can document actual grower decision making in the context of ecosystem processes and agronomic practices (National Research Council, 1989b; Shennan et al., 1991). To gain additional information, researchers can design complementary experiments to control these variable agronomic practices (Shennan et al., 1991). The complex interactions among farming system components warrants further investigation of pest-management implementation in whole-farming systems.

Moving from the discovery phase to the implementation phase requires information on scale-up, compatibility of new and established technologies, and other factors that effect the viability of approaches in commercial agriculture. Some components of private industry can efficiently move new products or procedures from small-scale tests to farm-level demonstrations of efficacy. Small industries, however, such as certain producers of predatory and parasitic arthropods and microbials, do not have the resources to conduct implementation research. Furthermore, pest-control practices that are not product-related do not have private-sector support, although funding for the discovery of these practices through competitive funding agencies is often available. Investments in discovery research should be accompanied by investments in the implementation of new discoveries. Increased attention to this area of research should result not only in increased implementation, but also in directions for discovery research.

Evaluation of the Impact of New Technologies

It can be argued that a concerted evaluation of the early impact of the chemical-control paradigm on factors such as agricultural and economic long-term viability, environmental contamination, and human health would have provided warnings of potential problems. Implementation of EBPM strategies must be tied to studies of the effect these new methods may have on nontarget species, agricultural practices, and human health.

Research to Improve Understanding of the Socioeconomic Issues Affecting Adoption

Social and economic factors will play significant roles in determining whether EBPM will be widely adopted. Research is needed to (a) develop management systems that are both economically and technically feasible, (b) target research and development at problems where solutions have the greatest value, (c) develop methods that can prolong the economic life of EBPM, and (d) provide the foundation for public policies that facilitate the development and implementation of EBPM.

The social and economic research needed can be divided into three broad categories:

• methods for measuring the direct and indirect effects of pest management systems for use in benefit-cost accounting;
• interdisciplinary research to compare the economic benefits and biological performances of different EBPM systems; and
• clarification of factors that influence risk-taking by agricultural producers.

Methods for Measuring Direct and Indirect Effects

It is now apparent that benefit-cost analyses of EBPM systems, including their initial and long-term social benefits and costs, should be factored into setting priorities and developing public policies for agricultural production. Such analyses will allow a rigorous comparison of alternatives and provide decision makers with the full range of costs and benefits of EBPM. It also will provide a systematic way of comparing one approach to another as well as provide a common framework for the analysis of alternatives.

Benefit-cost analyses should include an account of both direct and indirect costs and benefits of alternative management systems. Some direct benefits and costs will involve comparative cost of pesticide replacement versus the cost of the biological-control agent, the application costs, and effects of cultural practices that may be necessary to enhance the effectiveness of the treatment. Indirect benefits and costs involve

- human health effects to workers and consumers;
- environmental effects of the two practices on soil, water, and air quality;
- effects on nonpests, including wildlife and beneficial organisms;
- limitations on land use, such as reduced flexibility in crop rotation and crop selection;
  - establishment of other pests;
  - development of resistance by pests to the control procedures; and
  - lack of reliability of the management practices.

Although the direct costs of alternative agriculture systems may be greater than a chemically based system, indirect costs of the alternative system may be relatively low. Research is needed to identify the appropriate variables in comparing pest management approaches, and methods must be developed to collect comparative data. This will involve integrating the work of natural and social scientists, two groups that traditionally have not worked together. Such cooperation between disciplines, however, is important to developing cost-benefit analyses because many of the benefits of EBPM are indirect but essential to responsible ecosystem management. At the current level of understanding, if all indirect costs and benefits were considered in decisions concerning adoption of pest-control practices, benefit-cost analysis probably would indicate a significant advantage for EBPM.

Research to Compare Economic Benefits and Biological Performance

Economic feasibility is the most important factor to producers considering alternative management systems. Unfortunately, economic feasibility studies of the systems comprising EBPM are limited. Studies that have been done are often too narrow, ignoring decisions at the grower level and the impact of commodity prices, or estimating the profitability of the ecologically based management system without considering the profitability of traditional systems (Reichelderfer, 1981). Studies of the economic feasibility of ecologically based management systems relative to conventional systems need to be done for a variety of crops and geographic areas. Understanding the relative economic performance of this type of management system is essential to adoption and will focus future research and development on which aspects of the management system should be optimized.

The lack of systematically collected economic and performance data comparing the various pest-control methods continues to limit cost-benefit analyses. In 1979, because data on the degree of pest suppression and yield for the evaluated practices were not available, Reichelderfer (1979) was forced to conclude that all the pest-control methods evaluated achieved the same degree of pest control and produced the same crop yield. Similar problems with the lack of data constrained evaluation of different control methods for soybean cyst nematode (Reis et al., 1983) and the navel orangeworm, *Amyelois transitella* (Headley, 1983).

Clarification of Factors that Influence Risk Taking

Risk, and the way producers manage it, is an important factor in determining the speed of adoption of new technologies. Few studies have attempted to measure or otherwise quantify growers' attitudes toward the risk of using EBPM; however, Antle (1987) showed that the attitudes of populations and individuals toward risk can be estimated. Knowing the distribution of risk attitudes in a population of producers could help refine strategies for how new pest management systems should be presented.

Development of New Institutional Approaches to Encourage the Necessary Interdisciplinary Cooperation

The difficulty of manipulating the ecology of a cropping system to reduce losses caused by pests must not be underestimated. Every crop faces multiple assaults from numerous arthropods, diseases, and weeds, simultaneously; thus management decisions become complex. Many of the chemical and cultural tactics (e.g., tillage and crop rotation) available to agricultural and forest producers have broad impacts on the system. In suppressing one pest, it is likely that other components of the system will be disrupted. Research into the interactions among organisms in managed systems will enable scientists to design tools and methodologies to reduce populations of problematic organisms without negatively affecting the balance of the system.

The complexity of managed ecosystems necessitates coordinated multidisciplinary and interdisciplinary research to develop and implement EBPM. Funding for single-investigator initiated research projects will continue to be important, but for the most part single-investigator research will not sufficiently address the multidisciplinary nature of biological-control research. It is, therefore, essential that research institutions, and research policy, facilitate multidisciplinary and cooperative research. Frequently institutions, traditions, and policies erect barriers to interdisciplinary research. Efforts must be made to explore the nature of these barriers and to pull them down.

Multidisciplinary Research

Despite the generally recognized need for a systems approach to solving pest problems, the operating philosophy and organizational infrastructure in pest management is largely fragmented between different disciplines, agencies, and research institutions. This fragmentation is partly a result of the tendency for science to solve a problem by breaking it into parts—i.e., disciplines. Science education and research experiences are largely clustered around subject disciplines. For example, a biologist may specialize in molecular and cellular processes, or in whole organisms and organism-organism interactions, or in analyzing patterns that govern whole ecosystems. It is rare that a single researcher is accomplished in research across all these levels of biological organization. Research expertise can be further segmented by types of organisms and functions such as virology, entomology, plant pathology, microbiology, plant physiology, plant breeding, taxonomy, genetics, and epidemiology, just to list a few. Other disciplines are likewise divided into distinct groups that often consider each other rivals in the pursuit of resources and recognition.

Barriers have also been erected between the various pest science disciplines despite the fact that these groups are, in essence, interdisciplinary. Entomology, plant pathology, nematology, and weed science study the interactions of different groups of organisms and thus require broad backgrounds in the plant sciences, microbiology or invertebrate biology, and environmental sciences. The barriers are best exemplified by the disciplnes' development of parallel nomenclatures that describe similar processes, effectively further insulating the disciplines from each other.

In the agricultural sciences, clustering of research, extension, agribusiness, and agricultural policy around commodities hinders the development of broad solutions to problems. Frequently, funds for problem-solving research are provided by specific commodity groups, and research is restricted to the study of a single crop rather than the cropping system as a whole. Institutional structures, including professional societies and academic departments, and consequently funding patterns for research, are largely responsible for the development of barriers that then become hardened through competition for limited research and extension funds and other types of institutional support and recognition. Events in the past few decades that are unique to the pest sciences are challenging efforts to encourage interdisciplinary cooperation between these groups.

IPM programs are the intellectual antecedents to EBPM. As are EBPM strategies, IPM programs were developed on the basis of reducing pesticide use and integrating all pest-management strategies into a coherent unit. Competition for the limited funds, and the way that these funds were administered, led to rifts between and within the pest-science disciplines. Many programs that were conducted under the rubric of IPM can be criticized for becoming "*insect* pest management" programs or "integrated *pesticide* management" programs. It is unfortunate that because of the way IPM programs were implemented, one of IPM's legacies is more, rather than fewer, barriers between the pest-science disciplines.

The goals of EBPM cannot be reached without collaboration across taxon-specific disciplines and across hierarchical levels of organization. Interdisciplinary grants and other incentives for collaboration among groups must be crafted in ways that make researchers accountable for doing truly collaborative work. Interdisciplinary activity in the pest sciences must contend with historically based divisions at the research and administration levels or the problems that plagued IPM research will be repeated.

Professional Societies

The pest science professional societies, such as the Entomological Society of America, American Phytopathological Society, Society of Nematologists, and the Weed Science Society of America, each have members who advocate increased emphasis on biological controls for management of pests; but because these societies have diverse interests, they are not likely to provide the necessary leadership for EBPM. A strong, unified group that encourages multidisciplinary research and provides a forum for discussions of common themes related to EBPM could help to assure future interdisciplinary approaches to biological control. Forums for research and extension interactions between the pest science disciplines can also reverse the evolution toward different nomenclatures describing common phenomena. It is time for a professional forum to enhance communication among all scientists involved in aspects of EBPM.

Coordination of Federal Departments and Agencies

Ecologically based pest management is a topic of interest to a number of federal departments and agencies, and strong federal support is needed if progress is to be made in its implementation. The need for a coalition of credible advocates for EBPM from the research and extension communities is essential; strong leadership at the federal level can help to ensure that efforts in research and extension at the state and local levels are supported and coordinated.

A successful model for coordinating problem-solving research is the National Institutes of Health (NIH). NIH works to provide a national research agenda and funding to support both basic and applied research toward solving specific problems. NIH also acts to focus public attention on the problems being addressed. This type of model has been recommended in the past for agricultural research.

Attempts have been made within USDA to coordinate biological-control activities; this type of effort needs to be encouraged and expanded to include other federal departments and agencies with common interests. Identifying a single group within USDA empowered to speak for the various interests within the department and coordinate their activities would facilitate research and extension efforts in EBPM; this group could also coordinate UDSA's activities with those of other government departments and agencies with interests in this area, such as the Environmental Protection Agency, the National Science Foundation, the Department of Energy, the Agency for International Development, and the Department of the Interior.

One issue that will necessitate federal leadership is that of the overlapping goals—reduction of the use of pesticides—of both EBPM and IPM. The primary differences between EBPM and IPM are (a) the domination of pesticides in IPM and (b) IPM's history of implementation primarily for control of arthropods. When EBPM research proves successful, IPM programs will be the transition between current methods and EBPM. Unfortunately, because of the overlapping goals of the two programs, it may be that, rather than working together, there will be competition between the two approaches for limited resources and recognition. Federal leadership is needed to prevent this problem.

INFRASTRUCTURE FOR RESEARCH

EBPM will not be widely adopted without significant new research funding being made available to address current limitations to discovery, development, and implementation. Significant expansion of research in this area faces obstacles posed by financial and political stresses on local agricultural and natural resource research institutions. Even during the best financial times, institutions are slow to change research directions and methods. The down-sizing that is occurring in many research institutions is constraining the ability of administrators to redirect resources into new programs. This is particularly true when EBPM research competes with widely accepted conventional pest-control strategies, many of which are supported by the pesticide industry. It is cheaper to run trials of conventional pesticides to determine their utility in the local area than it is to identify and optimize application strategies for EBPM.

The need for large local research investments in EBPM research is analogous to the need that led to past investments in animal and plant breeding programs. Initial successes in plant breeding efforts demonstrated how directed breeding could greatly improve crops and domestic animals. Nevertheless, it took, and continues to take, a major investment in research to reap the benefits for agriculture from genetics. Confidence in a lucrative return drove this major investment even though the return often came decades after the initial investment was made. There have been enough successes now in biological-control research to justify a similar research investment in EBPM.

The research infrastructure needed to support a significant EBPM initiative exists both at the state and federal levels. The extensive USDA-ARS and land grant university research systems provide research personnel and facilities that are necessary to develop and implement EBPM. However, EBPM research must compete for resources with other research priorities. Major shifts in resource allocations will be necessary before a significant number of scientists become involved in research leading to EBPM.

Providing adequate research funding to accelerate the development and implementation of EBPM is difficult because of the diversity of research needs. Progress in biological control, for example, has been greatest where the necessary basic information is available or easily obtained. Progress has been most rapid at the macroscopic level where population ecology and predator-prey models are understood. Progress at the microscopic level has been slow because microbial ecology is still relatively poorly understood.

Research must also extend beyond basic to applied research. The need for applied research is compounded by the site-specificity of many ecologically based approaches. Ecologically based management systems with universal applicability are rare. Site specificity places unusual research, development, and extension burdens on local institutions. EBPM strategies adaptable to local conditions for each arthropod, weed, or pathogen pest are needed. Sharing research responsibilities between states can help reduce costs, but each locality will need to independently identify and optimize procedures for each of the major pests of the region.

The base of ecological information necessary to develop and implement EBPM is much greater than that for conventional chemical pesticides. Widespread implementation of EBPM will come only as a result of political processes through which the public makes it known that alternatives to conventional pesticides must be found.

Public Oversight of Ecologically Based Pest Management

The release of living predatory, parasitic, pathogenic, or antagonistic organisms (biological-control organisms), the deployment of biologically derived products such as toxins or semiochemicals (biological-control products), and the planting of resistant crop varieties (resistant plants) are fundamental components of EBPM. Wide-scale implementation of EBPM could include thousands of commercialized biological-control organisms and products, each of which could be quite specific with respect to target pest or disease as well as agronomic setting (area and crop) to which it would be applied. Any EBPM approach may entail risk to human health or to the environment. Appropriate oversight, therefore, is required to ensure that potential risks are properly assessed and managed, thereby promoting public acceptance of the use of biological-control organisms, biological-control products, or resistant plants. However, it is essential that the cost of meeting oversight requirements should not unduly constrain the development and implementation of promising methods of EBPM. In assessing risk, oversight by public agencies must use appropriate criteria and methods to avoid delays and unnecessary duplication.

An effective oversight system is essential for accelerating the development of EBPM. Such a system will quickly and adequately evaluate human and environmental health risk while considering the benefits of the new biological-control strategy. Construction of an effective oversight process presents a challenge for the future. Many specific biological-control organisms and products may be required to implement EBPM. To date, 43 microbial biological-control organisms have been registered by EPA.

Effective oversight requires risk assessment criteria suited to evaluating any potentially harmful effects of ecologically based controls on human and environmental health. Generally speaking, the risks posed by biological-control organisms and resistant plants are not the same as those associated with broad-spectrum synthetic chemical controls. The specificity of the mode of action of the biological organism, or resistant plant with respect to target pest, the likely constrained geographic area of biologically based control measures, and the method of deployment all contribute to the likelihood of safety. Human and environmental health risks of biological controls are not only different in kind from those of most conventional chemicals but also lower in the degree of hazard potentially posed. In many cases, the risks of continued use of chemical controls may be greater than the risks of instituting ecologically based strategies. By the same token, for pest problems not currently addressed by conventional chemicals, the use of ecologically based controls may indeed present some risks, but to the best of this committee's knowledge there needs to be more data to make a more accurate risk assessment. Nevertheless, the history of biologically based resistance genes supports their continued use and their risks may well be acceptable compared to those attributable to lack of effective control.

Biological-control products include genes or gene products derived from living organisms that kill, disable, or otherwise regulate the behavior of living organisms. In this category of controls, the product of a living organism rather than the living organism itself is used to manage a pest or pathogen population. Assessment of the potential risks of biological-control products, as chemical pesticides, must include their effects on environmental and human health. The nature of these risks is in large part a function of the mode of action of biological molecules, which range widely from toxicity to attraction. There are many natural products, including Bt toxins, pheromones, floral attractants, insect growth regulators, and plant growth regulators that have been registered as biochemical pesticides (U.S. Environmental Protection Agency, 1994a). Many others are being proposed for exemption from regulation by EPA. Fermentation products and plant extracts with a toxic mode of action have promise, but their development may be hindered if they are automatically regulated as conventional synthetic chemicals.

Effective oversight must also account for the likelihood that the particular kinds of human or environmental risks posed will vary with the biological-control organism, product, or resistant plant. And, the risks may vary in degree if not in kind depending on the particular agronomic setting of the pest/crop combination. Consequently, human health considerations may be paramount in some cases, whereas environmental effects (e.g., adverse impacts on nontarget organisms) may be the only cause for concern in other cases. Accommodating such variation in risk profiles across the many types of ecologically based controls will be key to constructing effective but not burdensome oversight.

Significant markets for agrochemicals in many cropping systems justify investment in the regulatory costs involved in their development, and some biological technologies similarly will have sufficient markets and profit potential to justify developers' covering the costs of complying with oversight requirements. However, pesticides used on minor crops or for limited purposes have less market potential. Some of the most efficacious and least risky products and approaches in EBPM will be modest in scale of use. Examples include release of classical self-perpetuating biological-control organisms that may produce great public good, but have virtually no commercial viability in the private sector. There is a need for a mechanism such as the IR-4 Program to encourage public support for development of biological-control organisms and products.

HUMAN HEALTH RISKS

Humans will come into contact with biological-control organisms, biological-control products, and resistant plants during their production and application and through exposure to organisms or residues that persist on crops and in the environment. The acute and chronic toxicities associated with conventional chemical pesticides are not found with most biological-control organisms and products. However, there may be other adverse health effects associated with their use. These include allergenic and other immune reactions and development of hypersensitivity through multiple exposure.

Health considerations are particularly relevant to workers involved in research on and production of biological controls. Workers applying or releasing the biological-control organisms, biological-control products, or resistant plants will also be exposed; therefore, packaging and application methods must address human exposure. EPA does require that hypersensitivity incidents be reported and may impose restrictions such as protective clothing to address potential problems associated with applicator exposure. It may be that adverse effects of biological-control organisms on human health could be restricted to certain individuals. For example, sensitivity to arthropod proteins is experienced only by some people who work extensively in arthropod rearing. Such risks can be managed by screening personnel for allergenicity or hypersensitivity and by providing protection against excessive exposure. When large-scale production of a biological-control organism is undertaken, concern about human exposure to proteins or other reactive materials is further warranted because sustained exposure to these materials can lead to acquired sensitivity and associated adverse reactions.

Microbes can produce numerous toxic and noxious metabolites, and mycotoxin contamination of food and feed from certain microbial control organisms is a legitimate concern (Betz et al., 1990). Appropriate screening for toxic components should therefore be included in risk assessment. In many cases, however, metabolites produced by biological-control agents are highly specific to the target pest and pose a minimal threat to human health. For example, Bt toxins are quickly degraded within the mammalian gut, reducing the potential for their accumulation within the mammalian system. Infrequently, the proteins cause mild skin irritation in allergic individuals. If, as is typical, the Bt toxin is applied on crops until the day of harvest, there is a probability of residue present in or on edible portions of the treated crop, but these residues present little risk because of the low mammalian toxicity (Panetta, 1993). Acute toxicity data should be obtained about the toxins produced by macrobes or microbes used in ecologically based pest-management approaches in order to characterize the relative risks they pose. In the event that acute toxicity is observed in these initial tests, it may then be appropriate to conduct chronic testing on the toxin itself.

As with conventional pesticides, residues of biological-control organisms or products may persist; for example, low numbers of pests and their biological-control organisms may be found in the harvested crop. Alternatively, resistant plant cultivars may pose risks to consumers if the edible portion of the crop is altered in harmful ways. The acute and chronic toxicities associated with broad-spectrum conventional pesticides may be less likely with biological-control organisms and products and resistant plants. This lower level of toxicity in biological controls may be attributed to a nontoxic mode of action, a mode of action for which there is no counterpart in mammals, low application rate, low potential for persistence and bioaccumulation of natural molecules, and low potential for applicator exposure to high concentrations of a pesticidal substance (except in the case of purified, concentrated extracts or fermentation products).

Host plant resistance can be achieved by decreasing or deleting the plant's production of a chemical that attracts a pest or is a nutrient or vitamin required for pest growth; alternatively, resistance can be achieved by incorporating genes that produce substances that are toxic to, repel, or inhibit the growth of pests. In cases in which there is also a chemical change in the edible part of the crop plant, there is a potential for effects on human nutrition and health. There have been cases in which a resistant commercialized cultivar led to illness in people who consumed the product or harvested the crop, caused by elevated levels of specific toxic compounds. If new or higher levels of potentially toxic proteins or secondary metabolites are produced, then acute or chronic testing may be indicated.

The emphasis of genetic engineers is on altering the plant to express single proteins that are either toxic to pests or inhibit reproduction of the pest. But there is potential for engineering of more complex traits—for example, secondary biochemical pathways with altered levels of certain compounds or biosynthesis of novel compounds. If either scenario exists and the competitiveness of the recipient organism changes, then additional studies are required to examine impacts on human health and the ecosystem.

It is certainly possible that new compounds produced in plants could also be toxic to nontarget organisms at low concentrations. For example, the Bt genes currently in use by genetic engineers encode a toxin that is effective at 1 microgram per gram tissue or less. Most plant-produced toxins are less effective than the Bt toxin at low concentrations (Leemans et al., 1990).

It is not clear that toxic resistance factors produced through genetic engineering pose a greater risk than secondary plant compounds produced through traditional breeding. These secondary compounds often have broad-based toxicity; and the effects of this toxicity on human health are poorly understood. More important, the biochemical basis for the resistance achieved through traditional plant breeding is often unknown. Understanding the mechanism of the resistance in genetically engineered plants can provide more confidence that the resistant plant will not adversely affect ecosystems or human health.

ENVIRONMENTAL RISKS

Conventional chemical pesticides may adversely affect groups of nontarget, economically desirable organisms, such as pollinators, predaceous insects, fish, and other wildlife (Higley and Wintersteen, 1992). Broadcast chemical sprays contact nontargets as well as targeted organisms, both inside and outside of the treatment area as a result of drift and runoff of the pesticidal materials. Such broad impact has generated pressure to move to less disruptive management practices (Pimentel et al., 1980). The scale for potential harm is generally very much smaller when considering nontarget effects of biological-control organisms.

Adverse effects on nontarget (nonpest) organisms and creation of new pests are the risks to environmental health that may be presented by EBPM strategies. Use of biological-control organisms and products as well as of resistant plants potentially can harm nontarget organism populations directly or indirectly though altered interactions within the pest complex. Use of control organisms and resistant plants can result in a new pathogen or weed through genetic exchange with other microbes and plants (Betz et al., 1987; Higley and Wintersteen, 1992; Hollander, 1991; Howarth, 1983, 1991; Hoy, 1992; Hoy et al., 1991; Lockwood, 1993; Nafus, 1993; National Audubon Society, 1994; Pimentel et al., 1992; Tiedje et al., 1989; U.S. Department of Agriculture, Agricultural Biotechnology Research Advisory Committee, 1991). The kinds and severity of these risks vary with the organism, product, or plant and the environmental and agronomic setting of the pest-management problem. For example, the length of time an organism is expected to exert effects on the pest/host system will help determine its opportunities to harm nontarget organisms. Similarly, the host range of a biological-control organism will determine the set of nontarget organisms it might attack. Fortunately, there is a considerable knowledge base on which to determine these manifestations of risks (Alexander, 1990; Caltagirone and Huffaker, 1980; Charudattan, 1990a; Cook, 1993; Ehler, 1990; Howarth, 1991; Hoy, 1992; Lockwood, 1993; National Research Council, 1989a; Tiedje et al., 1989).

Nontarget Effects

With biological-control macroorganisms, the most commonly cited risk is the potential for the organism to extirpate (eliminate through disruption) populations of nontarget organisms. Extirpation of nontarget species might result from predation, parasitism, pathogenicity, competition, or other attack on the nontarget species by the organism. This risk is most significant when a macroorganism is exotic (a nonnative species imported from outside the target area) and potentially not influenced by ecological controls to which an organism native to the area would be subjected. Still, the history of introductions of macroorganisms as biological-control organisms indicates a low incidence of problems (Clausen, 1978). The most notable examples of adverse nontarget effects from the use of macroorganisms as biological-control organisms involve the introduction of vertebrates such as the cane toad and mongoose (DeBach, 1974) and predatory snails (Howarth, 1991). However, many of these cases occurred in restricted or island populations, where habitat limitations appeared to have imposed constraints on the establishment of long-term balance between the introduced organism and its host/prey.

Biological-control organisms can attack less favored organisms if the target organism becomes scarce (Howarth, 1983, 1991). Local populations of organisms closely related taxonomically and ecologically to the target of the biological-control organism are most at risk to nontarget effects. Although an introduced organism may parasitize or prey on a species closely related to the target (Clausen, 1978; DeLoach, 1978; Krombein et al., 1979; Wapshere, 1982), this is not of serious concern unless extinction of the nontarget population is possible (Howarth, 1991). Host-switching exhibited by a biological-control organism often involves attack on a nontarget host or prey that falls within the organism's natural range of suitable hosts and thus does not necessarily involve a change in behavior or genetic adaptation by the organism. Even with some elasticity in host range or habitat range, there are limits to host/prey switching that confer a level of specificity to the use of macroorganisms as controls and lessen the risk of nontarget effects (Harris, 1988; Watson, 1985). Even after thorough investigation of the physiological host range of a new biological-control organism, and with data on ecological host ranges of closely related organisms, additional assessments of affected hosts must be made after the organism is introduced into the field. Although there is virtually no empirical evidence of harmful prey switching by exotic biological-control organisms, few thorough studies have been reported (National Audubon Society, 1994).

Experimentation can provide useful insights into the impact of changes in host-range specificity of pathogens over time. The potential for changes in host range is an important influence in assessing the risk of using pathogenic biological-control organisms. Typically with an exotic pathogen, a specific pathotype is identified for introduction and studied in its native environment to delineate its host range and pathogenic potential. With an indigenous pathogen, a specific isolate or pathotype may be used initially, but additional selection may be necessary to identify a more effective strain. In this situation, the host-range specificity of the original pathotype may not represent that of the selected strain. Hence, it is advisable to assess the host range of each improved pathotype or strain.

Hypotheses regarding the innate ability of natural enemies to parasitize or prey on nontarget species can be tested under microcosmic conditions in a laboratory or greenhouse. But the physiological host range revealed by such experiments may not represent the true (ecological) host range of the organism (Olckers et al., 1995; Watson, 1985). Single-choice experiments (one organism with one host) are poorly discriminatory and are likely to give false-positive results. Nonetheless, they are useful in eliminating concern about potential nontarget hosts. Multiple-choice experiments (one organism with two or more different hosts) reduce the chances of false-positives; nevertheless, the ecological host range expressed under field conditions is likely to be much more narrow than the physiological host range characterized by laboratory choice tests (Ridings et al., 1978; TeBeest, 1988; Watson, 1985). Any prediction of the ecological host range must consider both the experimentally determined physiological host range and the biological, ecological, and taxonomic host-range of any closely related organisms.

The prediction of ecological host range is particularly difficult with self-perpetuating organisms that function at the tertiary trophic level, such as predatory and parasitic arthropods. Although it can be determined in laboratory conditions whether a predatory or parasitic arthropod will utilize a range of hosts, such data cannot be freely extrapolated to field conditions. Complex spatial, temporal, and behavioral interactions commonly displayed by self-perpetuating macro-organisms ultimately determine whether an organism has even the opportunity to encounter a given host.

After release of the organisms into the field, appropriate monitoring will be necessary to check the reliability of the physiological host-range data. Monitoring over several years will identify direct and indirect effects on natural communities. Every introduction provides the opportunity to validate and/or modify protocols for estimating ecological specificity from microcosm studies. Climate, geography, floral phenology, and trophic interactions are key aspects that combine to define the relationship of an organism with coexisting organisms. Thus, ecological host range can be estimated with some accuracy, combining physiological host-range data with ecological evidence from other geographical areas where a similar system (ecological analogy) is in place, particularly if closely related biological-control organisms have previously been monitored there after release (National Audubon Society, 1994).

In general terms, a threshold level of a pest is required to sustain reproduction of a predatory or parasitic biological-control organism; and if high numbers of the biological-control organism suppress the capacity of the pest to reproduce, then negative feedback adversely affects the survival of the biological-control organism (National Audubon Society, 1994). As pest population levels decline, there is a decreased likelihood of contact between the biological-control organism and the pest, resulting in reduced reproduction of the biological-control organism. Although this traditional theory of predator-prey relationship remains valid, other components such as foraging behavior also contribute to these ecological interactions (National Research Council, 1986b). Current knowledge of host-prey relationships is useful as a starting point in assessing risk of nontarget effects on macroorganisms. However, the continuing contributions of ecologists, biologists, entomologists, weed scientists, botanists, zoologists, and other scientists will be necessary for a comprehensive evaluation of these complex ecological systems.

Because of the enormous diversity of microbial pathogens that might be used as pest controls, there is incomplete understanding of their potential adverse effects on nonpathogenic microorganisms in natural systems. Because microorganisms occupy numerous soil or plant habitats, laboratory experiments can provide data only on selected populations, and evaluation of nontarget effects from these studies may not be applicable to agricultural ecosystems (Hollander, 1991; Tiedje et al., 1989). Though the species composition of any particular microbial ecosystem may not be known, there is considerable knowledge about some of the functional roles and mechanisms of microbes found there; this information can be used to identify potential effects on nontarget species (Cook, 1991; National Research Council, 1989a; Tiedje et al., 1989).

Certain microorganisms, such as *B. thuringiensis,* suppress plant pests by producing toxins or antibiotics, and risk evaluation of these biological-control organisms is focused on the persistence of the toxin and its effects on nontarget species. Experience has demonstrated a basis for concern about potential effects of δ-endotoxin (Bt) on nontarget arthropods within the class against which the biological-control product is active, particularly in the case of certain endangered species of butterflies. In these cases, the habitats of concern have been identified and product use has been prohibited there (Hutton, 1992). This approach is especially appropriate when a gene encoding a toxin is introduced into an organism that occupies a habitat where the toxin would not otherwise be found. The encapsulated Bt protein toxin is nontoxic to vertebrates and has been demonstrated to be of minimal risk to other nontarget organisms (Panetta, 1993); it has established an excellent precedent, but similar benign properties must be established for other microbial toxins that may be useful in EBPM. Molecular techniques can provide scientists with the tools to make precise alterations in toxin-encoding genes, thereby improving product performance and increasing the ability to focus the effect of the control on the target organism.

The long history of plant breeding suggests that resistant plant cultivars rarely cause significant effects on nontarget organisms. Plant resistance factors range from chemical to physical and can be specific to a single pest species or have broad effects on an array of organisms as different as arthropods and pathogens.

The potential risks to nontarget organisms resulting from new or elevated levels of toxins in resistant plant cultivars are not well known. If a cultivar resistance is toxicity based, then there is potential for native herbivores such as caterpillars or honey bees to be harmed, though these occurrences are not well

Collego®: A Mycoherbicide Approach for Weed Management

COLLEGO® is one of two commercial bioherbicide products commercially available since the early 1980s. COLLEGO® is a formulation of the spores of a fungal plant pathogen *Colletotrichum gloeosporiodes* [Penz.] Sacc. f.sp. *aeschynomene* (Coelomycetes). The product was developed by scientists of the Arkansas Agricultural Experiment Station at the University of Arkansas; the U.S. Department of Agriculture's Agricultural Research Service's Rice Research Station at Stuttgart, Arkansas; and The Upjohn Company. COLLEGO® is used to manage northern jointvetch, *Aeschynomene virginica* (L.) B.S.P., a native leguminous weed in rice and soybean crops in Arkansas, Mississippi, and Louisiana.

The fungus produces an anthracnose disease that can kill both seedling and mature northern jointvetch plants. The natural concentrations of the fungus found in fields are generally inadequate to control northern jointvetch. Effective weed control can be achieved, however, by augmenting natural populations of the fungus by spraying fields with a suspension of fungal spores. The fungus can be easily grown in culture and produces spores abundantly, making commercial formulation easy. COLLEGO® is a wettable powder of dried spores sold in three components: (1) the spore powder, (2) a hydrating liquid, and (3) activated charcoal to clean spray tanks. It is applied using aerial or ground sprayers after the crop has emerged, preferably soon after rain or irrigation.

COLLEGO® has consistently provided more than 90 percent control of northern jointvetch. Although the fungus has a much broader host range than originally thought (it can infect several nontarget legumes including English peas [*Pisum sativum*]), it has not posed any danger to nontarget plants under field conditions during nearly 2 decades of experimental and commercial use.

COLLEGO® has served as a model for bioherbicide science, technology, and regulation. Since its commercial introduction in 1982, it has been used repeatedly and successfully, performed consistently, been integrated with other pest management and cultural practices, proved highly stable in its virulence, and produced excellent scientific information garnered from a decade of follow-up research.

SOURCE: TeBeest, D. O., and G. E. Templeton. 1985. Mycoherbicides: Progress in the biological control of weeds. Plant Dis. 69:6-10.

---

known. Additional field and laboratory studies can improve an assessment of such risks.

Some resistance factors are expressed only at one stage of the crop's growth—for example, resistance to second-generation corn borer (Brindley et al., 1975). Some are expressed in only one organ of the plant—for example, in the silks of corn—and some are expressed only when the plant is under attack by the pest or pathogen (Ryan, 1983). Other resistance factors are present in all stages of growth and in most organs of the plant. The range of species affected by a plant resistance factor is an indicator of potential effects on nontarget organisms. This is analogous to host range and specificity determining the potential effects of a biological-control organism on nontarget organisms. Superior resistance factors are those that decrease multiple pest or pathogen populations without adversely affecting nontarget organisms (Kennedy and Barbour, 1992). However, there are cases in which improved resistance to one pest decreases resistance to others (Dacosta and Jones, 1971), which could be a severe limitation to product efficacy in some situations.

In some cases, host plant resistance changes the plant in a way that benefits natural enemies of a pest (Price et al., 1980). For example, the production of extrafloral nectaries or changes in leaf architecture may enhance the ability of a biological-control organism to find its target pest organism (Schuster and Calderon, 1986). Certain cultivars of wheat are thought to be resistant to a soil-borne disease as a result of enhanced populations of antagonists. However, host-plant resistance can adversely affect biological-control organisms as well as pests (Robb and Bradley, 1968) by reducing populations of arthropod hosts parasitized or preyed on by control organisms. There are no clear examples where negative effects on natural enemies have overridden the positive effects of pest reduction. Applied entomology has a history of assessing potential negative effects on beneficial arthropods before resistant cultivars are released.

Exacerbation of Plant Pests

Assessment of the safety of a biological-control organism introduced into the environment involves several factors, including familiarity with the organism and its function, prior history of use, and characteristics of the target environment. Much has been learned from prior releases of biological-control organisms that can provide a basis for assessment. If the organism is unfamiliar or there is uncertainty about the environment into which it is introduced, a careful evaluation must be conducted prior to introduction into the environment (National Research Council, 1989a).

If there is a risk associated with introducing biological-control organisms, products, and cultivars into the environment, it is the low probability of producing a pest that previously was not a pest. Microorganisms have been determined to exchange genetic information in the environment, but with low frequency. Transfer of genetic information may result in (a) acquisition of virulence or (b) enhancement of host range or virulence. Although such enhancement may contribute to enhanced pathogenicity or compromise the durability of biological tools introduced into the environment, the pathogenicity of a microorganism is known to result from a complex interaction among a number of genes and gene products of the pathogen and host. Hence the pathogenic potential of a microbe can be anticipated.

The probability of a nonpathogenic species becoming a virulent pathogen is extremely low. To confer pathogenicity, a pathogen must first attach itself to a suitable host, compete for nutrients, and also, in the case of saprophytes, resist the defense systems of the host. It is highly unlikely that moving one or a few genes from a pathogen to an unrelated nonpathogen will cause the recipient to become pathogenic. Species most at risk are those that are most closely related to the pathogen, while those less closely related must overcome considerable genetic barriers (National Research Council, 1989a).

Some documentation does exist to confirm gene transfer between microorganisms. Resistance genes of some introduced microorganisms are transferred to indigenous microorganisms such as *A. tumefaciens*. Cisar and colleagues demonstrated an enhancement of host range or virulence of the mycoherbicide Collego® through transfer of genetic material to a related indigenous pathogen (Cisar et al., 1994). Pathogens with enhanced host range or virulence may be generated through genetic instability; for example, if rearrangements occur in the genome of disabled viroids placed into transgenic plants, there is potential for generation of virulent forms (Hammond, 1994). Some other unknown consequences of genetic exchange may exist, but comprehensive field evaluation can demonstrate occurrence of these events in a natural setting.

Weeds may result from crosses between introduced and native plants. Wild plants have been brought into the United States from other countries for breeding purposes. These wild plants have been crossed with cultivated crop varieties in open-field tests often with few precautions against the development of new weeds. Origins of weeds can be traced to (a) wild colonizers, (b) hybridization between wild and cultivated races of domestic species, and (c) abandoned domesticated varieties (Oka and Morishima, 1982). Experience with intensive crop breeding during this century indicates that hybridization between wild relatives and germplasm used for crop improvement is relatively rare.

The experience with sorghum is quite instructive and can be used as a general model for understanding the potential for development of weed-crop complexes and new pernicious weeds. According to De Wet (1966), the weediness of johnsongrass (*Sorghum halepense*) was enhanced coincidentally with its introgression with cultivated sorghums (*Sorghum bicolor*) in the United States. When these johnsongrass populations extend their already major ecological role outside agricultural fields, they represent the most extreme category of known risk associated with gene flow from a crop to a weedy relative.

Other examples of gene exchange (gene flow or introgression) between a domesticated crop and its wild relative has been reported for cultivated rice (*Oryza sativa*) and wild rice (*Oryza* spp.) (Parker and Dean, 1976). Gene exchange between corn and its wild relative teosinte, between Eastern carrots and wild carrots, and between durum wheat and wild emmer wheat have been documented. The genus *Amaranthus* is prone to natural hybridization between the few cultivated species and several weedy relatives (Simmonds, 1979). In some cases, the hybrids between the cultivated and wild amaranths could out-compete their weed parents (Tucker and Sauer, 1958).

Biological Control of Crown Gall

Agrobacterium radiobacter strain K84 is a naturally occurring bacterium that controls crown gall, a plant tumor caused by the related soil bacterium *A. tumefaciens*. Crop losses caused by crown gall occur worldwide and can be extensive, particularly in nurseries growing rosaceous plants, grapevines, and stone fruit trees. *A. tumefaciens* enters the plants through wounds and causes tumors, which can weaken, reduce the aesthetic quality, and eventually kill the host plant. There are no effective chemical controls currently available for crown gall.

The biological control agent *A. radiobacter* strain K84 has been used commercially for more than a decade in many regions of the world, including Australia, Greece, Israel, Italy, Japan, New Zealand, South Africa, Spain, and the United States. K84 can be applied to wounds on cuttings, bare-rooted seedlings, grafts, and on field-grown plants. K84 protects wounds from infection by *A. tumefaciens* in part because of the production of agrocin 84, an antibiotic with specific toxicity against sensitive strains of *A. tumefaciens*.

Genes determining the production of agrocin 84 and immunity of the host bacterium to agrocin 84 are present on pAgK84, an indigenous, conjugative plasmid of strain K84. If plasmid pAgK84 is transferred to *A. tumefaciens* through the natural process of bacterial conjugation, the pathogen becomes immune to agrocin 84 and less sensitive to biological control by strain K84. In response to concerns that the predominance of *A. tumefaciens* harboring pAgK84 may reduce the efficacy of biological control, a derivative strain of K84, lacking a region (*tra*) required for conjugal transfer of pAgK84, has been constructed (Jones et al., 1988). This strain cannot transfer pAgK84 to other bacteria and its use is expected to minimize the risk that biological control will break down due to the presence of agrocin 84-resistant strains of *A. tumefaciens* in nursery soils. The strain containing the *tra* deletion is used commercially in Australia.

Gene flow has apparently occurred from cultivated rye (*Secale cereale* spp.) to wild relatives in California, where a weedy rye probably derived from a cross between *S. cereale* and *S. montanum* (wild relative) has become increasingly crop-like. This introgression has proceeded to such an extent that farmers are said to have abandoned efforts to grow cultivated rye for human consumption; instead they deliberately sow hybrids for forage (Jain, 1977; Suneson et al., 1969).

Although hybridization between a crop and its wild relative may not be preventable, there is little likelihood that desirable domesticated traits will be retained in the wild relative. Much of the emphasis in plant breeding has been on traits that would reduce adaptation to the wild. Important commercial traits, such as pest resistance, that have the potential to alter the ecology of wild relatives have not been a problem, with the possible exception of gene transfer from cultivated sorghum to johnsongrass (National Research Council, 1989a). In general, experience with traditionally bred resistant plants suggests that although the potential for genetic transfer between resistant varieties and wild relatives exists, it occurs infrequently. Where it has occurred, it caused a problem only rarely because the weed became less weedy and more domesticated. Still, there is need for vigilance in the search for evidence of negative effects of resistance-gene transfer (National Research Council, 1989a).

RISK ASSESSMENT AND MANAGEMENT

Evaluation of risks associated with deployment of biological-control organisms and products and resistant plants should be based on evidence relative to both the type of organism or product and its method of deployment.

EBPM involves the deployment of a range of tactics from application of synthetic chemical pesticides to liberation and establishment of live organisms. This range of tactics might be viewed as a continuum, with application of synthetic chemical pesticides on one end and relocation of living self-perpetuating arthropods or microbes on the other. Along the continuum one might encounter "natural" pesticides derived from plants (sabadilla, rotenone) or microbes (Bt δ-endotoxin, others) that are deployed using methods and expectations similar to those used for synthetic pesticides. Further yet along the continuum one could encounter living, nonperpetuating and self-perpetuating organisms whose effects may be derived by the in situ production of toxins or other deleterious factors. Among these are entomopathogenic viruses and plants containing pest or disease resistance capabilities. It should not be difficult to envision, then, that the processes, fundamental principles, and expectations of deployment of tactics along this "continuum" would vary widely from one portion of the continuum to another. Therefore, basing the conceptual or empirical framework for oversight of all organisms or products on characteristics, risk factors, and history of one could lead to inappropriate risk evaluation and failure to meet the goal of ensuring human and environmental safety.

Drawing on Experience and Experimentation

All experience including that gained from the natural occurrence of the biological materials and their toxicity and field testing should be fully considered in regulatory review of new organisms or products.

There are innumerable combinations of controls (organisms, products, resistant plants), pests (pathogens, weeds, arthropods), and agroecological settings (geographic location, crop). If attempted on a case-by-case basis, requiring experimentation unique to each, oversight requirements and costs could well become prohibitive, particularly given the small market size of many biological controls. Fortunately, the task of risk assessment for one situation can be greatly expedited by drawing on relevant experience with similar controls, pest, and agroecologies. Knowledge gained from past releases or uses of related organisms or products is the best guide for evaluating the potential risks and benefits of new releases of organisms in the same category. As experience grows with related taxa or functionally similar organisms, adjustments can be made in the categorization and oversight required for new organisms. For example, USDA's Animal and Plant Health Inspection Service (APHIS) proposed that introduction of certain categories of genetically modified plants would require notification rather than a permit (Federal Register, 1992). The EPA proposed exemption from the notification requirement for certain subgroups of microbial pesticides, as information warranting such action becomes available (U.S. Environmental Protection Agency, 1994b).

For any particular biological-control organism or product or resistant plant, expeditious risk assessment will rely on information generated by field experience with closely related organisms or substances and/or by appropriate experimentation in the laboratory or greenhouse. Consequently, the protocol for human and environmental risk assessment will vary accordingly. Relevant risks must be assessed, the information or additional inquiries needed to support risk assessment must be identified and evaluated, and a synthesis of the findings must be made: but by whom? For EBPM, the range of expertise and experience required to make such judgments transcends that which public agency professional scientific staff can reasonably be expected to cover. Consequently, public oversight must employ expert review committees consisting of scientists from relevant disciplines in agriculture, ecology, and human health as well as practitioners with experience with closely related organisms or products. Although public officials remain the ultimate arbiters of the acceptability of any risks posed by a new pest control, the diverse nature of biological-controls dictates involvement of scientists with relevant experience or expertise simply on the grounds of efficiency, avoiding imposition of unnecessary or duplicate information requirements.

Setting Priorities

Experience, experimentation, and expert opinion should direct oversight attention to broad-spectrum organisms and products or resistant plants and their use on major acreage crops where risk factors are greatest or most difficult to assess. At the same time, effective review will exempt or remove from oversight those organisms, products, or resistant plants for which accumulated experience indicates low risk. Generally speaking, higher human and/or environmental risks will be associated with larger scale in geographic use and/or with the duration of effect of the control organism, product, or resistant plant. Conventional broad-spectrum pest control chemicals often pose risks because of persistence in the environment or as residues in food or because of wide use.

Scale of Use

The significance of scale of use of a pest control is already recognized, for example, in the differential oversight treatment by EPA of microbial control organisms depending on whether the proposed release is for small-scale testing (less than 10 acres), large-scale testing, or full commercial product registration (widespread use). Many biological-control organisms, even if fully commercialized would be limited to relatively small geographical areas because of their environmental or host specificity. Limitation in scale of use also limits human health risk because fewer individuals may be exposed in production or application of the control and because any dietary exposure would be restricted. Environmental risk is similarly restricted because the geographic specificity is related to the fact that the control is not useful or cannot survive in other agroecological settings.

Persistence

The persistence of a control organism may also affect the possibilities for adverse effects on the environment or increases in human exposure. Biological-control organisms can be

- self-perpetuating organisms that become permanently established;
- self-perpetuating organisms that can reproduce through one or multiple generations, but will ultimately expire (for reasons such as climatic extremes— e.g., mealybug destroyer in southern California not able to overwinter); or
- organisms that are incapable of self-perpetuation in the environments into which they are introduced.

An established self-perpetuating organism exerts permanent effects on the local native populations, the pest-pathogen complex, and on the gene pool. If any of these effects are potentially negative, the environmental risk of release is magnified by self-perpetuation.

The characteristics that confer a biological-control organism with the ability to self-perpetuate are

- tolerance of the environment and habitat,
- ability to compete with established populations of indigenous organisms,
- appropriate life-cycle synchrony with a target host (particularly important for macroorganisms), and
- ability to reproduce when host density is high.

Self-perpetuation requires the ability to survive during periods of low host abundance, to tolerate winter and other adverse conditions, or to survive by assuming a dormant form. In some cases, the likelihood of self-perpetuation then can be deduced from the taxon to which the organism belongs; in addition, the method and timing of deployment can influence establishment of the organism and its longevity.

With a nonperpetuating biological-control organism, site-confined application is possible. The fact that a nonperpetuating organism will not become a permanent component of an ecosystem may substantially reduce risk where there is a concern about human health and environmental effects (Howarth, 1991).

Microbial pathogens of plants can persist in nature. If susceptible hosts are present at adequate densities and environmental conditions are conducive, microbial pathogens will spread from the initial focus of establishment. When populations of the target host are reduced, then dispersal of the microbial pathogen slows and its numbers decline. Over time, the system assumes equilibrium, or homeostasis, when the control agent becomes endemic. Fitting this model are the postrelease histories of the following weed-control agents:

1. *Puccinia chondrollina*, a rust fungus from the Mediterranean region used to control skeleton weed (*Chondrilla juncea*) in Australia (Cullen, 1985) and California (Supkoff et al., 1988);

2. *Entyloma compositarum*, a smut fungus from Jamaica used to control hamakua pa-makani (*Ageratina riparia*) in Hawaii (Trujillo et al., 1988);

3. *Phragmidium violaceum*, a rust fungus from Europe used to control wild blackberries (*Rubus constrictus* and *R. constrictus*) in Chile (Oehrens, 1977); and

4. *Puccinia carduorum*, a rust fungus from Turkey used to control musk thistle (*Carduus thoermeri*) in Virginia (Baudoin et al., 1993).

Long duration of effect, by itself, is not a risk factor; the risk depends on the nature of the control organism. If an organism is host specific, for example, longevity will pose little risk to nontarget organisms. Indeed, the primary effect of longevity will be to increase the benefits of the organism as a pest or disease-control system. If the organism is not host specific, its longevity will increase the potential for effects on nontarget organisms. A persistent organism can become a pest or a nuisance in its new environment (Caltagirone, 1981; Harris, 1988; Howarth, 1983), but it is possible to minimize such risk. Phylogenetic, ecological, and biological relationships are indicative of the host ranges of related groups of biological-control organisms. When empirical evidence indicates a restricted host specificity, then risk from introduction of an organism from that group can be estimated (National Audubon Society, 1994).

Managing Risk

Conventionally, risk assessment is taken to be a separate exercise from risk management (National Research Council, 1983). As applied to human health concerns, risk assessment involves hazard identification (Does the agent cause the adverse effect?), dose-response assessment (What is the relationship between dose and effect in humans?), and exposure assessment (What exposures are currently experienced or anticipated under different conditions?). The resulting risk characterization answers the question, What is the estimated incidence of the adverse effect in a given population? Risk assessment is performed by public- or private-sector analysts in support of public agency decisions about risk management and acceptable levels of risk. In the case of human health effects, especially with respect to carcinogens, legislative mandate dictates the tolerable level of risk.

Although levels of acceptable risk are established in the public arena, not by the scientific community, guidance should be sought from experience in managing risks similar to those posed by biological-control organisms and products or resistant plants. For example, the Food and Drug Administration's current view of the health risks of genetically engineered plants (that the fact of their modification does not a priori raise concerns) is instructive and demonstrates that the requirements of public oversight need not be burdensome. Environmental risks posed to nontarget organisms, particularly when extinction might be a possibility, have been managed through the registration process for conventional chemicals. Determination of the acceptable risk should be made without regard to the process that produced the modified organism (i.e., conventional breeding versus genetic engineering), recognizing the relevance of experience already gained and the precision of newer techniques (National Research Council, 1989a).

Gaps and Inconsistencies in Current Oversight

The current oversight responsibility for biological-control organisms, products and resistant plants is shared primarily between EPA and USDA/APHIS and is characterized by application of different oversight requirements for microorganisms versus macroorganisms, for exotic versus indigenous organisms, and for organisms considered to be plant pests and those not considered plant pests. The complexities and anomalies of the current system may be attributed to the overlapping jurisdiction of several agencies, the diversity of organisms to be regulated, and the attempt to make the decision making "template" developed for registration of conventional chemical pesticides applicable to biologically based controls.

Current federal oversight generally evaluates the risks of biological-control macroorganisms using the procedures developed for exotic plant pests under the Federal Plant Pest Act (FPPA); microbial biological-control agents are subject to oversight procedures developed for chemical pesticides under the Federal Insecticide, Fungicide, Rodenticide Act (FIFRA). Neither set of procedures is tailored for registration of biological-control agents; therefore, both pose unnecessary barriers to registration of biological-control organisms.

The goal of FPPA is to prevent the introduction or interstate movement of exotic plant pests. Although the statute does not specifically provide for comprehensive regulation of all biological-control organisms, all such organisms (even those that are only capable of attacking pest arthropods) are assumed by APHIS to be potential plant pests and therefore subject to regulation under FPPA.

FIFRA was intended for regulation of chemical pesticides and poses severe constraints to the efficient testing of microbial biological-control organisms. Once separated from exotic or endemic plant pests under quarantine or similar containment conditions, biological-control organisms should be subject to risk assessment protocols commensurate with their biological and ecological attributes.

Although there are cases in which the pesticide "template" functions effectively for large-scale commercial registration of biological-control products, they do not function efficiently for registration of biological-control organisms. Neither the intent of the biological-control organism application nor the process by which it operates parallels the chemical pesticide model.

A case in point is the convoluted process required for the registration of encapsulated Bt-endotoxins (a killed microbial pest control agent that was engineered to produce the Bt toxin). Killed microbial products produced through the fermentation of a live intermediate are further regulated by EPA under the Toxic Substances Control Act (TSCA) as pesticidal intermediates, despite the fact that such agents are contained within a fermentor until the final killed microbial pesticide is produced. Such evaluation includes a review of the potential health and environmental effects of the inadvertent release of the live intermediate. This situation results in a dual review process within EPA, with the manufacturing portion of the process reviewed separately from the field release of the killed microbial pesticide. This is apparently the result of the dual jurisdictions of FIFRA and TSCA for review of "chemical intermediates," a distinction that is irrelevant to microbial pesticides.

The existing fragmented oversight process employs different assessment criteria to biological-control organisms that pose similar human or environmental risks, resulting in unnecessary scrutiny of some low-risk organisms and almost complete lack of regulatory review of others. The most important inconsistency is the differential treatment of macroorganisms by USDA/APHIS under FPPA and of microorganisms by EPA under FIFRA. For example, neither EPA nor APHIS regulates nonpathogenic nematodes, whereas both agencies regulate microorganisms. In addition, microbes are exempt from oversight if vectored by a nonpathogenic, indigenous macroorganism (such as a nematode). There is no clearly defined process for risk assessment of biological-control macroorganisms that are not plant pests, resulting in indecisiveness in granting permits for field release. The current regulatory oversight of macroorganisms is thus reduced to case-by-case evaluations through environmental assessments or environmental impact statements under the National Environmental Protection Act (NEPA). To provide appropriate oversight of biological-control organisms and products, USDA, EPA, and other federal and state agencies with oversight authority should jointly develop criteria and protocols specifically for the testing, registration, and use of living control organisms, biological-control products, or plants resistant to pests or diseases.

No formal oversight structure exists for traditionally bred resistant plants, whereas genetically engineered resistance will, it appears, be subject to close scrutiny (U.S. Environmental Protection Agency, 1994a). A question confronting policy makers is whether host plant resistance accomplished through genetic engineering should receive a higher degree of scrutiny than resistance achieved through traditional plant breeding. Such a distinction imbedded in regulatory practice would not necessarily reflect differences between actual human or environmental health risks posed by the two technologies.

Options for Improvement

The barriers to effective, efficient oversight for biological-control organisms and products and resistant plants are inconsistencies in the existing framework of laws and regulations and the resultant overlap or lack of coordination among agencies with jurisdiction. In fashioning a more coherent approach to assessing the human and environmental risks of EBPM, there is a need to depart from current practice. Most important, the criteria for risk assessment must vary with the possibilities presented by each organism, product, or resistant plant.

To accommodate the demand for information based on experience as well as experimentation, oversight must draw on the resources of the scientific community and of field practitioners. Technical Advisory Groups (TAG), a consortium of representatives from government and research established to provide advice to the government on the safety of biological organisms for weed control, may be useful for oversight of supplements of EBPM.

The U.S. Environmental Protection Agency and the U.S. Department of Agriculture, currently responsible for oversight of pesticide regulations, should develop and publish a guide to risk criteria, data requirements, and oversight procedures that apply to importation, movement, introduction, testing, and release or registration of biological-control organisms or products.

A regulatory road map would be a first and important step toward reducing the uncertainties and delays caused by gaps and inconsistencies in current regulatory treatment of biological-control organisms. Such a road map would be particularly valuable now in the early stages of development of a broad-based industry producing biological-control organisms and products and resistant plants. Potential developers of new controls especially those in academic settings may be unfamiliar with oversight requirements and procedures, and an available road map would reduce the costs of acquiring that knowledge. As discussed previously, the costs of complying with oversight requirements can be an important determinant of commercial viability of biological pest control with small potential markets. Even today, the slow pace of registration may be at least partially attributable to the significant costs of gaining regulatory approval.

Streamlining the oversight process would enable those involved in research, development, and use of biological-control organisms and products to anticipate the requirements of public evaluation of product efficacy and human and environmental health risks. Those involved should be able to identify, through an integrated road map or similar mechanism, the agencies with authority for the specific organism or product to be evaluated. The road map should also provide detailed guidance through the procedures and requirements. Where procedures have associated timetables, these should be spelled out. Through the development of a logistical or procedural road map, an individual or group would be able to identify the full process at the outset, and would also be able to assess the data requirements, criteria for evaluation, and costs. In addition, an estimate of the time period necessary for completion of the evaluation process would be available to assist in planning and prioritization of projects.

This committee visualizes EBPM as the foundation for an approach to not only managing but also to assuring the durability of biological-control organisms, biological-control products, and resistant cultivars. Principles of ecology that lay the foundation of EBPM must be incorporated into implementation and oversight. An important goal of EBPM is to restore and preserve balance to the managed ecosystem by duplicating natural processes to the maximum extent possible. Risk assessment also should reflect that principle. Biological-control products must be developed and implemented in ways that complement managed ecosystems and facilitate the biological and natural controls already existing to suppress pests. Monitoring of new products and processes is key. The resultant information will lead to early identification of durability problems. The knowledge gained from this monitoring of the dynamic interactions among organisms will increase the understanding needed to manage old and new pests in a safe, profitable, and durable way.

References

Adams, C. E. 1994. The role of IPM in a safe, healthy, plentiful food supply. Pp. 25–34 in Proceedings of the Second National Integrated Pest Management Symposium/Workshop. Durham: North Carolina State University.

Alabouvette, C. 1993. Naturally occurring disease-suppressive soils. Pp. 204–210 in Pest Management: Biologically Based Technologies: Proceedings of the Beltsville Symposium XVII, R. D. Lumsden and J. L. Vaughn, eds. Washington, D.C.: American Chemical Society.

Alexander, M. 1990. Potential impact on community function. Pp. 121–125 in Risk Assessment in Agricultural Biotechnology: Proceedings of the International Conference, J.J. Marois and G. Bruening, eds. Oakland: Division of Natural Resources, University of California.

Alms, M. J. 1994. Challenges for IPM in the future: An independent crop consultant's perspective. Pp. 81–84 in Second National Integrated Pest Management Symposium/Workshop: Proceedings. April 19–22, 1994. Las Vegas, Nevada.

Andres, L. A., and F. D. Bennett. 1975. Biological control of aquatic weeds. Annu. Rev. Entomol. 20:31–46.

Andrews, J. H. 1991. Comparative Ecology of Microorganisms and Macroorganisms. New York: Springer-Verlag.

Antle, J. M. 1987. Econometric estimation of producers' risk attitudes. Am. J. Agric. Econ. 69:509–522.

Baker, D. B., and R. P. Richards. 1989. Herbicide concentration patterns in rivers draining intensively cultivated farmlands of northwestern Ohio. Pp. 103–119 in Pesticides in Terrestrial and Aquatic Environments, D. Weigmann, ed. Blacksburg, Va.: Virginia Water Resources Research Center, Virginia Polytechnic Institute and State University.

Baudoin, A. B. A. M., R. G. Abad, L. T. Kok, and W. L. Bruckart. 1993. Field evaluation of *Puccinia carduorum* for biological control of musk thistle. Biol. Control 3:53–60.

Betz, F., A. Rispin, and W. Schneider. 1987. Biotechnology products related to agriculture: Overview of regulatory decisions at the U.S. Environmental Protection Agency. Pp. 316–327 in Biotechnology in Agricultural Chemistry, H. M. LeBaron, R. O. Mumma, R. C. Honeycutt, J. H. Duesing, J. F. Phillips, and M. J. Haas, eds. Washington, D.C.: American Chemical Society.

Betz, F. S., S. F. Forsyth, and W. E. Stewart. 1990. Registration requirements and safety considerations for microbial pest control agents. Pp. 3–10 in Safety of Bicrobial Insecticides, M. Laird, L. A. Lacey, and E. W. Davidson, eds. Boca Raton, Fla.: CRC Press.

Bottrell, D. G. 1979. Integrated Pest Management Council on Environmental Quality. Washington, D.C.: U.S. Government Printing Office.

Brazzel, J. R. 1989. Boll weevil eradication—An update. Pp. 218–220 in Proceedings of the Beltwide Cotton Conference, Book 1. Memphis, Tenn.: National Cotton Council of America.

Brazzel, J. R., T. B. Davich, and L. D. Harris. 1961. A new approach to boll weevil control. J. Econ. Entomol. 54:723–730.

Brindley, T. A., A. N. Sparks, W. B. Showers, and W. D. Guthrie. 1975. Recent research advances on the European corn borer in North America. Annu. Rev. Entomol. 20:221–239.

Caltagirone, L. E. 1981. Landmark examples in classical biological control. Annu. Rev. Entomol. 26:213–232.

Caltagirone, L. E., and C. B. Huffaker. 1980. Benefits and risks of using predators and parasites for controlling pests. Ecol. Bull. 31:103–109.

Carlson, G. A. 1988. Economics of biological control of pests. Am. J. Alternative Agric. 3:110–116.

Carlson, G. A., G. Sappie, and M. Hammig. 1989. Economic returns to boll weevil eradication. Econ. Res. Serv. Rep. 621:1–31.

Carson, R. 1962. Silent Spring. Boston: Houghton Mifflin.

Cate, J. R. 1988. Population management of boll weevil in sustainable cotton production systems. Pp. 249–254 in the Proceedings of the Beltwide Cotton Producers and Researchers Conference. Memphis, Tenn.: National Cotton Council and The Cotton Foundation.

Cate, J. R., and M. K. Hinkle. 1993. Integrated Pest Management: The Path of a Paradigm. Washington, D.C.: National Audubon Society.

Charudattan, R. 1986. Integrated control of water hyacinth (*Eichhornia crassipes*) with a pathogen, insects, and herbicides. Weed Sci. 34 (Suppl. 1): 26–30.

Charudattan, R. 1990a. Release of fungi: Large-scale use of fungi as biological weed control agents. Pp. 70–84 in Risk Assessment in Agricultural Biotechnology: Proceedings of the International Conference, Publ. No. 1928, J. J. Marois and G. Bruening, eds. Oakland, Calif.: University of California, Division of Agriculture and Natural Resources.

Charudattan, R. 1990b. Biological control of aquatic weeds by means of fungi. Pp. 186–201 in Aquatic Weeds: The Ecology and Management of Nuisance Aquatic Vegetation, A. H. Pieterse and K. J. Murphy, eds. New York: Oxford University Press.

Charudattan, R. 1990c. Microbial control of aquatic weeds. Pp. 71–78 in Proceedings of European Weed Research Society Eighth Symposium on Aquatic Weeds, P. R. F. Barrett, M. P. Greaves, K. J. Murphy, A. H. Pieterse, P. M. Wade, and M. Wallsten, eds. The Netherlands: Wageningen.

Charudattan, R., J. T. DeValerio, and V. J. Prange. 1990. Special problems associated with aquatic weed control. Pp. 287–303 in New Directions in Biological Control: Alternatives for Suppressing Agricultural Pests and Diseases, R. Baker and P. E. Dunn, eds. New York: Alan R. Liss.

Cisar, C. R., F. W. Spiegel, D. O. TeBeest, and C. Trout. 1994. Evidence for mating between isolates of *Colletotrichum gloeosporidoides* with different host specificities. Curr. Gen. 25:330–335.

Clausen, C. P. 1978. Introduced Parasites and Predators of Arthropod Pests and Weeds: A World Review. Agricultural Handbook No. 480. Washington, D.C.: U.S. Department of Agriculture.

Cook, R. J. 1990. Twenty-five years of progress toward biological control. Pp. 1–14 in Biological Control of Soil-Borne Plant Pathogens, D. Hornby, ed. Wallingford, U.K.: CAB International.

Cook, R. J. 1991. Biological Control of Plant Diseases: Broad Concepts and Applications, Technology Bulletin No. 123. Taipei City, Republic of China on Taiwan: Food and Fertilizer Technology Center.

Cook, R. J. 1993. Making greater use of introduced microorganisms for biological control of plant pathogens. Annu. Rev. Phytopathol. 31:53–80.

Cook, R. J., and K. F. Baker. 1983. The nature and practice of biological control of plant pathogens. St. Paul, Minn.: American Phytopathological Society.

Cook, R. J., and D. M. Weller. 1987. Management of take-all in consecutive crops of wheat or barley. Pp. 41–76 in Innovative Approaches to Plant Disease Control, I. Chet, ed. New York: Wiley-Interscience.

Cook, R. J., C. J. Gabriel, A. Kelman, S. Tolin, and A. K. Vidaver. 1995. Research on plant disease and pest management is essential to sustainable agriculture. BioScience 45:354–357.

Cooksey, D. A. 1990. Genetics of bactericide resistance in plant pathogenic bacteria. Annu. Rev. Phytopathol. 28:201–219.

Costa, A. S., and G. W. Müller. 1980. Tristeza control by cross protection: A U.S.-Brazil cooperative success. Plant Dis. 64:538–541.

Council on Environmental Quality. 1972. Integrated Pest Management. Washington, D.C.: U.S. Government Printing Office.

Cullen, J. M. 1985. Bringing the cost benefit analysis of biological control of *Chondrilla juncea* up to date. Pp. 145–152 in Proceedings of the VI International Symposium on Biological Control of Weeds, E. S. Delfosse, ed. Ottawa: Canadian Government Publications Centre.

Dacosta, C. P., and C. M. Jones. 1971. Cucumber resistance and mite susceptibility controlled by the bitter gene in *Cucumis sativus* L. Science 172:1145–1146.

Davis, J. R., O. C. Huisman, D. T. Westermann, L. H. Sorensen, A. T. Schneider, and J. C. Stark. 1994. The influence of cover crops on the suppression of Verticillium wilt of potato. Pp. 332–341 in Advances in Potato Pest Biology and Management. St. Paul, Minn.: APS Press.

DeBach, P. 1974. Biological Control by Natural Enemies. New York: Cambridge University Press.

DeBach, P., and M. Rose. 1977. Environmental upsets caused by chemical eradication. Calif. Agric. 31:8–10.

DeBach, P., and D. Rosen. 1991. Biological Control by Natural Enemies. Cambridge, U.K.: Cambridge University Press.

Dean, H. A., J. V. French, and D. Meyerdick. 1983. Development of integrated pest management in Texas citrus. Bull. Texas Agric. Exp. Stat. B-1434.

Dekker, J. 1993. The fungicide resistance problem: Current status and the role of systemics. Pp. 163–180 in Pesticide Interactions in Crop Production: Beneficial and Deleterious Effects, J. Altman, ed. Boca Raton, Fla.: CRC Press.

DeLoach, C. J. 1978. Considerations in introducing foreign biotic agents to control native weeds of rangelands. Pp. 39–50 in Proceedings of the IV International Symposium on Biological Control of Weeds, T. E. Freeman, ed. Gainesville, Fla.: University of Florida.

De Wet, J. M. J. 1966. The origin of weediness in plants. Proc. Oklahoma Acad. Sci. 47:14–17.

Doutt, R. L., and R. F. Smith. 1971. The pesticide syndrome: Diagnosis and suggested prophylaxis. In Biological Control, C. B. Huffaker, ed. New York: Plenum Press.

Edwards, C. R. 1991. National organization promotes integrated pest management. Am. Entomol. 37:136–137.

Edwards, C. R., and R. E. Ford. 1992. Integrated pest management in the corn/soybean agroecosystem. Pp. 13–55 in Food, Crop Pests, and the Environment, F. G. Zalom and W. E. Fry, eds. St. Paul, Minn.: APS Press.

Ehler, L. E. 1990. Environmental impact of introduced biological control agents: Implications for agricultural biotechnology. Pp. 85–96 in Risk Assessment in Agricultural Biotechnology: Proceedings of the International Conference, J. J. Marois and G. Bruening, eds. Oakland: Division of Natural Resources, University of California.

Federal Register. 1992. 57(Nov. 6/216):53036–53043.

Fernandez-Cornejo, J., E. D. Beach, and W.-Y. Huang. 1992. The Adoption of Integrated Pest Management Technologies by Vegetable Growers. Washington, D.C.: U.S. Department of Agriculture, Economic Research Service, Resources and Technology Division.

Ferris, H. 1992. Biological approaches to the management of plant-parasitic nematodes. Pp. 68–101 in Beyond Pesticides: Biological Approaches to Pest Management in California, T. Beall, ed. Oakland, Calif.: University of California.

Ferro, D. N. 1993. Integrated pest management in vegetables in Massachusetts. Pp. 95–105 in Successful Implementation of Integrated Pest Management for Agricultural Crops, A. R. Leslie and G. W. Cuperus, eds. Boca Raton, Fla.: Lewis Publishers.

Fitchen, J. H., and R. N. Beachy. 1993. Genetically engineered protection against viruses in transgenic plants. Annu. Rev. Microbiol. 47:739–763.

Fitzner, M. S. 1993. The role of education in the transfer of biological control technologies. Pp. 382–387 in Pest Management: Biologically Based Technologies, R. D. Lumsden and J. L. Vaughn, eds. Washington, D.C.: American Chemical Society.

Flint, M. L. 1992. Biological approaches to the management of arthropods. Pp. 2–30 in Beyond Pesticides: Biological Approaches to Pest Management in California, T. Beall, ed. Oakland, Calif.: University of California.

Flint, M. L., and R. van den Bosch. 1981. Introduction to Integrated Pest Management. New York: Plenum Press.

Frisbie, R. E. 1989. Critical issues facing IPM technology transfer. Pp. 157–162 in Proceedings of the National Integrated Pest Management Symposium/Workshop. Geneva, N.Y.: Communications Services, New York State Agricultural Experiment Station, Cornell University.

Frisbie, R. E., and J. W. Smith. 1989. Biologically intensive integrated pest management: The future. Pp. 151–164 in Progress and Perspectives for the 21st Century, J. J. Menn and A. L. Steinhauer, eds. Lanham, Md.: Entomological Society of America.

Frisbie, R. E., and G. M. McWhorter. 1986. Implementing a statewide pest management program for Texas, U.S.A. Pp. 234–262 in Advisory Work in Crop Pest and Disease Management, J. Palti and R. Ausher, eds. New York: Springer-Verlag.

Frisbie, R. E., D. D. Hardee, and L. T. Wilson. 1992. Biologically intensive integrated pest management: Future choices for cotton. Pp. 57–82 in Food, Crop Pests, and the Environment: The Need and Potential for Biologically Intensive Integrated Pest Management, F. G. Zalom and W. E. Fry, eds. St. Paul, Minn.: APS Press.

Gelvin, S. B. 1992. Chemical signaling between *Agrobacterium* and its plant host. In Molecular Signals in Plant-Microbe Communications, D. P. S. Verma, ed. Boca Raton, Fla.: CRC Press.

Georghiou, G. P. 1986. The magnitude of the resistance problem. Pp. 14–44 in Pesticide Resistance: Strategies and Tactics for Management. Washington, D.C.: National Academy Press.

Gerson, U., and E. Cohen. 1989. Resurgences of spider mites (Acari: Tetranychidae) induced by synthetic pyrethroids. Exp. Appl. Acarol. 6:29–46.

Goe, W. R., and M. Kenney. 1988. The political economy of the privatization of agricultural information: The case of the United States. Agric. Admin. Exten. 28:81–99.

Goeden, R. 1993. Arthropods for suppression of terrestrial weeds. Pp. 231–237 in Pest Management: Biologically Based Technologies: Proceedings of the Beltsville Symposium XVII, R. D. Lumsden and J. L. Vaughn, eds. Washington, D.C.: American Chemical Society.

Gould, F. 1991. The evolutionary potential of crop pests. Am. Sci. 79:496–507.

Graebner, L., D. S. Moreno, and J. L. Baritelle. 1984. The Fillmore Citrus Protective District: A success story in integrated pest management. Bull. Entomol. Soc. Am. 30:27–33.

Grieshop, J. I., F. G. Zalom, and G. Miyao. 1988. Adoption and diffusion of integrated pest management innovations in agriculture. Bull. Entomol. Soc. Am. 34:72–78.

Griffiths, E. 1993. Iatrogenic effects of pesticides on plant diseases—An update and overview. Pp. 269–279 in Pesticide Interactions in Crop Production: Beneficial and Deleterious Effects, J. Altman, ed. Boca Raton, Fla.: CRC Press.

Guzelian, P. S., C. J. Henry, and S. S. Olin, eds. 1992. Similarities and Differences Between Children and Adults: Implications for Risk Assessment. Washington, D.C.: International Life Sciences Institute Press.

Hall, R. W., and L. E. Ehler. 1979. Rate of establishment of natural enemies in classical biological control. Bull. Entomol. Soc. Am. 25:280–282.

Hall, R. W., L. E. Ehler, and B. Bisabri-Ershadi. 1980. Rate of success in classical biological control of arthropods. Bull. Entomol. Soc. Am. 26:111–114.

Hallberg, G. R. 1989. Pesticide pollution of groundwater in the humid United States. Agric. Ecosyst. Environ. 26:299–368.

Hammond, R. W. 1994. *Agrobacterium*-meditated inoculation of PSTVd cDNAs onto tomato reveals the biological effect of apparently lethal mutations. Virology 201:36–45.

Hanna, R., F. G. Zalom, and C. L. Elmore. 1995. Integrating cover crops into grapevine pest and nutrition management: The transition phase. Sustainable Agriculture Technical Reviews. USDA Sustainable Agriculture Research and Education Program. Washington, D.C.: U.S. Department of Agriculture.

Harley, K. L. S., and I. W. Forno. 1990. Biological control of aquatic weeds by means of arthropods. Pp. 177–186 in Aquatic Weeds: The Ecology and Management of Nuisance Aquatic Vegetation, A. H. Pieterse and K. J. Murphy, eds. New York: Oxford University Press.

Harris, P. 1988. Environmental impact of weed-control insects. BioScience 38:542–548.

Headly, J. C. 1983. Economic analysis of naval orangeworm control in almonds. Calif. Agric. 37(5/6):27–29.

Headly, J. C. 1985. Cost-benefit analysis: Defining research needs. Pp. 53–63 in Biological Control in Agricultural IPM Systems, M. A. Hoy and D. C. Herzog, eds. New York: Academic Press.

Higley, L. G., and W. K. Wintersteen. 1992. A novel approach to environmental risk assessment of pesticides as a basis for incorporating environmental costs into economic injury levels. Am. Entomol. 38:34–39.

Hirano, S. S., and C. D. Upper. 1990. Population biology and epidemiology of *Pseudomonas syringae*. Annu. Rev. Phytopathol. 28:155–177.

Holden, L. R., J. A. Graham, R. W. Whitmore, W. J. Alexander, R. W. Pratt, S. K. Liddle, and L. L. Piper. 1992. Results of the national alachlor well water survey. Environ. Sci. Technol. 26:935–943.

Hollander, A. K. 1991. Environment impacts of genetically engineered microbial and viral biocontrol agents. Pp. 251–266 in Biotechnology for Biological Control of Pests and Vectors, K. Maramorosch, ed. Boca Raton: CRC Press.

Holt, J. S., and H. M. LeBaron. 1990. Significance and distribution of herbicide resistance. Weed Technol. 4:141–149.

Howarth, F. G. 1983. Classical biocontrol: Panacea or Pandora's box. Proc. Hawaiian Entomol. Soc. 24:239–244.

Howarth, F. G. 1991. Environmental impacts of classical biological control. Annu. Rev. Entomol. 36:485–509.

Hoy, M. A. 1989. Integrating biological control into agricultural IPM systems: Reordering priorities. Pp. 41–57 in Proceedings of the National Integrated Pest Management Symposium/Workshop. Geneva, N.Y.: Communications Services, New York State Agricultural Experiment Station, Cornell University.

Hoy, M. A. 1992. Biological control of arthropods: Genetic engineering and environmental risks. Biol. Control 2:166–170.

Hoy, M. A., and D. C. Herzog. 1985. Biological Control in Agricultural IPM Systems. New York: Academic Press.

Hoy, M. A., R. M. Nowierski, M. W. Johnson, and J. L. Flexner. 1991. Issues and ethics in commercial releases of arthropod natural enemies. Am. Entomol. (Summer):74–75.

Huffaker, C. B., M. van den Vrie, and J. A. McMurtry. 1969. The ecology of tetranychid mites and their natural control. Annu. Rev. Entomol. 14:125–174.

Huffaker, C. B., M. van den Vrie, and J. A. McMurtry. 1970. The ecology of tetranychid mites and their natural enemies: A review. II. Tetranychid populations and their possible control by predators: An evaluation. Hilgardia 40:391–458.

Hutton, P. 1992. Regulation of microbial biological pest control agents by the Environmental Protection Agency. Pp. 25–29 in Regulations and Guidelines: Critical Issues in Biological Control, R. Charudattan and H. W. Browning, eds. Gainesville, Fla.: University of Florida, Institute of Food and Agricultural Science.

Ingels, C. 1995. Cover cropping in vineyards: A grower profile series, part 2. Am. Vineyard (June).

Jain, S. K. 1977. Genetic diversity of weedy rye populations in California. Crop Sci. 17:480–482.

Johal, G. S., and S. P. Briggs. 1992. Reductase activity encoded by the HM1 disease resistance gene in maize. Science 258:985–987.

Johnston, H. G. 1961. The impact of insecticidal resistance upon the use and development of insecticides for cotton pests. Pp. 41–44 in Entomology Society of America: Miscellaneous Publications, Vol. 2. College Park, Md.: Entomology Society of America.

Jones, A. L. 1982. Chemical control of phytopathogenic prokaryotes. Pp. 399–414 in Phytopathogenic Prokaryotes, Vol. 2, M. S. Mount, ed. Boca Raton, Fla.: CRC Press.

Jones, D. A., M. H. Ryder, B. G. Clare, S. K. Farrand, and A. Kerr. 1988. Construction of a Tra⁻ deletion mutant of pAgK84 to safeguard the biological control of crown gall. Mol. Gen. Genet. 212:207–214.

Kennedy, G. G., and J. D. Barbour. 1992. Resistance variation in natural and managed systems. Pp. 13–41 in Plant Resistance to Herbivores and Pathogens: Ecology, Evaluation, and Genetics, R. S. Fritz and E. L. Sims, eds. Chicago, Ill.: University of Chicago Press.

Kessmann, H., T. Straub, C. Hofmann, T. Maetzke, and J. Herzog. 1994. Induction of systemic acquired disease resistance in plants by chemicals. Annu. Rev. Phytopathol. 32:439–459.

Kim, D. G., and R. D. Riggs. 1994. Techniques for isolation and evaluation of fungal parasites of *Heterodera glycines*. J. Nematol. 26:592–595.

Knapp, J. L. 1995. 1995 Florida Citrus Pest Management Guide. Gainesville, Fla.: Florida Cooperative Extension Service.

Knipling, E. F. 1968. Boll weevil and pink bollworm eradication: Progress and Plans. Pp. 14–18 in Summary of the Proceedings of the 1968 Beltwide Cotton Conference at Hot Springs, Arkansas. Memphis, Tenn.: National Cotton Council of America.

Knipling, E. F. 1971. Technically feasible approaches to boll weevil eradication. Pp. 23–30 in Proceedings of the 1971 Beltwide Cotton Conference, Atlanta, Georgia. Memphis, Tenn.: National Cotton Council of America.

Kogan, M. 1986. Ecological Theory and Integrated Pest Management Practice. New York: John Wiley & Sons.

Kolter, R., D. A. Siegele, and A. Tormo. 1993. The stationary phase of the bacterial life cycle. Annu. Rev. Microbiol. 47:855–874.

Kramer, R. A., W. T. McSweeny, and R. W. Stavros. 1983. Soil conservation with uncertain revenues and input supplies. Am. J. Agric. Econ. 65:694–702.

Krombein, K. V., P. D. Hard, Jr., and B. D. Burks. 1979. Catalog of Hymenoptera in America North and Mexico, 3 volumes. Washington, D.C.: U.S. Government Printing Office.

Lawson, T. J. 1982. Information flow and crop protection decision making. Pp. 21–32 in Decision Making in the Practice of Crop Protection, R. B. Austin, ed. Croydon, U.K.: BCPC Publications.

Leemans, J., A. Reynaerts, H. Hofte, M. Peferoen, H. V. Mellaert, and H. Joos. 1990. Insecticidal crystal proteins from *Bacillus thuringiensis* and their use in transgenic crops. Pp. 573–581 in New Directions in Biological Control: Alternatives for Suppressing Agricultural Pests and Diseases. New York: Alan R. Liss.

Lewis, W. J., and W. R. Martin, Jr. 1990. Semiochemicals for use with parasitoids. J. Chem. Ecol. 16:3067–3089.

Lewis, W. J., L. E. M. Vet, J. H. Tumlinson, J. C. van Lenteren, and D. R. Papaj. 1990. Variations in parasitoid foraging behaviour. Essential element of a sound biological control theory. Environ. Entomol. 19:1183–1193.

Lin, B. H., M. Padgitt, L. Bull, H. Delvo, D. Shank, and H. Taylor. 1995. Pesticide and Fertilizer Use and Trends in U.S. Agriculture. Agricultural Economic Report No. 717. Washington, D.C.: Economic Research Service, U.S. Department of Agriculture.

Lindow, S. E. 1985. Integrated control and role of antibiosis in biological control of fireblight and frost injury. Pp. 83–225 in Biological Control on the Phylloplane, C. Windels and S. E. Lindow, eds. St. Paul, Minn.: APS Press.

Lindow, S. E. 1993. Biological control of plant frost injury: The ice story. Pp. 113–128 in Advanced Engineered Pesticides, L. Kim, ed. New York: Marcel Dekker.

Lindow, S. E. 1995. The use of reporter genes in the study of microbial ecology. Mol. Ecol. (in press).

Liss, W. J., L. J. Gut, P. H. Westigard, and C. E. Warren. 1986. Perspectives on arthropod community structure, organization, and development in agricultural crops. Annu. Rev. Entomol. 31:455–478.

Lockwood, J. A. 1993. Environmental impacts of classical biological control. Environ. Entomol. 22.

Loper, J. E., and S. E. Lindow. 1993. Roles of competition and antibiosis in suppression of plant diseases by bacterial biological control agents. Pp. 144–155 in Pest Management: Biologically Based Technologies, R. D. Lumsden and J. L. Vaughn, eds. Washington, D.C.: American Chemical Society.

Lumsden, R. D., J. A. Lewis, and J. C. Locke. 1993. Managing soilborne plant pathogens with fungal antagonists. Pp. 196–203 in Pest Management: Biologically Based Technologies, R. D. Lumsden and J. L. Vaughn, eds. Washington, D.C.: American Chemical Society.

Lumsden, R. D., J. A. Lewis, and D. R. Fravel. 1995. Formulation and delivery of biocontrol agents for use against soil-borne plant pathogens. Unpublished report.

Mahr, D. L. 1991. Implementing Biological Control in the North Central States: An Extension Perspective. Unpublished report.

Mahr, D. L. 1995. Biological Control in the United States: A Survey of the Perceptions, Resources, and Needs of Extension Entomologists. Unpublished report.

Martyn, R. D. 1985. Water hyacinth decline in Texas caused by *Cercospora piaropi*. J. Aquatic Plant Manag. 23:29–32.

McCoy, C. W., and T. L. Couch. 1982. Microbial control of the citrus rust mite with the mycoacaridice Mycar. Fla. Entomol. 65:115–126.

McMurty, J. A., C. G. Huffaker, and M. van den Vrie. 1970. Ecology of Tetranychid mites and their natural enemies: A review. I. Tetranychid enemies: Their biological characters and the impact of spray practices. Hilgardia 40:331–390.

Mellano, V. J., and D. A. Cooksey. 1988. Development of host-range mutants of *Xanthomonas campestris* pv. translucens. Appl. Env. Microbiol. 54:884–889.

Metcalf, R. L. 1982. Insecticides in pest management. Pp. 217–277 in Introduction to Insect Pest Management, 2nd Ed., R. L. Metcalf and W. H. Luckmann, eds. New York: Wiley & Sons.

Metcalf, R. L. 1983. Implications and prognosis of resistance to insecticides. Pp. 703–733 in Pest Resistance to Pesticides, G. P. Georghiou and T. Saito, eds. New York: Plenum.

Meyer, S., and R. Huettel. 1993. Fungi and fungus/bioregulator combinations for control of plant-parasitic nematodes. Pp. 214–221 in Pest Management: Biologically Based Technologies: Proceedings of the Beltsville Symposium XVII, R. D. Lumsden and J. L. Vaughn, eds. Washington, D.C.: American Chemical Society.

Moore, L. W., and G. Warren. 1979. *Agrobacterium radiobacter* strain 84 and biological control of crown gall. Annu. Rev. Phytopathol. 71:163–179.

Mumford, J. D. 1982. Farmers' perceptions and crop protection decision making. Pp. 13–19 in Decision Making in the Practice of Crop Protection, R. B. Austin, ed. Croydon, U.K.: BCPC Publications.

Murphy, K. J., and P. R. F. Barrett. 1990. Chemical control of aquatic weeds. Pp. 136–173 in Aquatic Weeds: The Ecology and Management of Nuisance Aquatic Vegetation, A. H. Pieterse and K. J. Murphy, eds. New York: Oxford University Press.

Nafus, D. M. 1993. Movement of introduced biological control agents onto nontarget butterflies, *Hypolimnas* spp. (Lepidoptera: Nymphalidae). Environ. Entomol. 22:265–272.

National Audubon Society. 1991. Insect Resistance to Bt δ-Endotoxin: What It Means for Farming Practices and the Environment. Washington, D.C.: National Audubon Society.

National Audubon Society. 1994. Host Specificity in Biological Control Organisms. Washington, D.C.: National Audubon Society.

National Research Council. 1969. Principles of Plant and Animal Pest Control, Part 3: Insect-Pest Management and Control. Washington, D.C.: National Academy Press.

National Research Council. 1975. Pest Control: An Assessment of Present and Alternative Technologies, Vol. 1, Contemporary Pest Control Practices and Prospects. Washington, D.C.: National Academy of Sciences.

National Research Council. 1983. Risk Assessment in the Federal Government: Managing the Process. Washington, D.C.: National Academy Press.

National Research Council. 1984. Genetic Engineering of Plants. Washington, D.C.: National Academy Press.

National Research Council. 1986a. Drinking Water and Health, Vol. 6. Washington, D.C.: National Academy Press.

National Research Council. 1986b. Ecological Knowledge and Environmental Problem-Solving. Washington, D.C.: National Academy Press.

National Research Council. 1986c. Pesticide Resistance: Strategies and Tactics for Management. Washington, D.C.: National Academy Press.

National Research Council. 1987. Regulating Pesticides in Food: The Delaney Paradox. Washington, D.C.: National Academy Press.

National Research Council. 1989a. Field Testing Genetically Modified Organisms. Washington, D.C.: National Academy Press.

National Research Council. 1989b. Alternative Agriculture. Washington, D.C.: National Academy Press.

National Research Council. 1993a. Groundwater Vulnerability Assessment. Washington, D.C.: National Academy Press.

National Research Council. 1993b. Pesticides in the Diets of Infants and Children. Washington, D.C.: National Academy Press.

National Research Council. 1993c. Soil and Water Quality: An Agenda for Agriculture. Washington, D.C.: National Academy Press.

National Research Council. 1994. Rangeland Health: New Methods to Classify, Inventory, and Monitor Rangelands. Washington, D.C.: National Academy Press.

Neate, S. M., and A. D. Rovira. 1993. Pesticide disease interactions in conservation tillage systems. Pp. 515–530 in Pesticide Interactions in Crop Production: Beneficial and Deleterious Effects, J. Altman, ed. Boca Raton, Fla.: CRC Press.

Nielsen, E. G., and L. K. Lee. 1987. The Magnitude and Costs of Groundwater Contamination from Agricultural Chemicals: A National Perspective. Agricultural Economic Report No. 576. Washington, D.C.: U.S. Department of Agriculture, Economic Research Service.

Norton, G. A. 1982. Crop protection decision making—An overview. Pp. 3–11 in Decision Making in the Practice of Crop Protection, R. B. Austin, ed. Croydon, U.K.: BCPC Publications.

Nuss, D. L. 1992. Biological control of chestnut blight: An example of virus-mediated attenuation of fungal pathogenesis. Microbiol. Rev. 56:561–576.

Oehrens, E. 1977. Biological control of the blackberry through the introduction of rust, *Phragmidium violaceum*, in Chile. FAO Plant Protect. Bull. 25:26–28.

Oka, H.-I., and Morishima, H. 1982. Ecological genetics and the evolution of weeds. Pp. 73–89 in Biology and Ecology of Weeds, W. Holzner and N. Numata, eds. The Hague, Netherlands: W. Junk Publishers.

Olckers, T., H. G. Zimmerman, and J. H. Hoffman. 1995. Interpreting ambiguous results of host-specificity tests in biological control of weeds: Assessment of two *Leptinotarsa* species (Chrysomelidae) for the control of *Solanum eleagnifolium* (Solanaceae) in South Africa. Biol. Control 5:(in press).

Oliver, J. D. 1993. Formation of viable but nonculturable cells. Pp. 239–272 in Starvation in Bacteria, S. Kjelleberg, ed. New York: Plenum Press.

Ollinger, M., and J. Fernandez-Cornejo. 1995. Regulation, Innovation, and Market Structure in the U.S. Pesticide Industry. Economic Research Service, Agricultural Economic Report No. 719. Washington, D.C.: U.S. Department of Agriculture.

Osteen, C. D., and P. I. Szmedra. 1989. Agricultural Pesticide Use Trends and Policy Issues. Agricultural Economic Report No. 622. Washington, D.C.: U.S. Department of Agriculture.

Panetta, J. D. 1993. Engineered microbes, the Cellcap® System. Pp. 379–392 in Advanced Engineered Pesticides, L. Kim, ed. New York: Marcel Dekker.

Parker, C., and M. L. Dean. 1976. Control of wild rice in rice. Pestic. Sci. 7:403–416.

Payne, C. C. 1988. Insect pest management concepts: The role of biological control. Pp. 1–7 in Biotechnology, Biological Pesticides, and Novel Plant-Pest Resistance for Insect Pest Management, D. W. Roberts and R. R. Granados, eds. Ithaca, N.Y.: Cornell University.

Pedigo, L. P., and H. G. Higley. 1992. The economic injury level concept and environmental quality: A new perspective. Am. Entomologist 38:12–21.

Pieterse, A. H. 1990. Biological control of aquatic weeds: Introduction to biological control of aquatic weeds. Pp. 174–177 in Aquatic Weeds: The Ecology and Management of Nuisance Aquatic Vegetation, A. H. Pieterse and K. J. Murphy, eds. New York: Oxford University Press.

Pimentel, D., D. Andow, R. Dyson-Hudson, D. Gallahan, S. Jacobson, M. Irish, S. Kroop, A. Moss, I. Schreiner, M. Shepard, T. Thompson, and B. Vinzant. 1980. Environmental and social costs of pesticides: A preliminary assessment. Oikos 34:126–140.

Pimentel, D., H. Acquay, M. Biltonen, P. Rice, M. Silva, J. Nelson, V. Lipner, S. Giordano, A. Horowitz, and M. D'Amore. 1992. Environmental and economic costs of pesticide use. BioScience 42:750–760.

Poehlman, J. M. 1979. Breeding Field Crops, Second Ed. Westport, Conn.: AVI Press.

Poehlman, J. M., and D. A. Sleper, eds. 1995. Breeding Field Crops, 4th edition. Ames: Iowa State University Press.

Popoff, F. P., and D. T. Berggelli. 1993. Chemical and Engineering News (January):8–10.

Poston, F. L. 1989. Extension priorities for IPM in the next decade. Pp. 163–165 in Proceedings, National Integrated Pest Management Symposium/Workshop. Geneva, N.Y.: Communications Services, New York State Agricultural Experiment Station, Cornell University.

Powell-Abel, P., R. S. Nelson, B. De, N. Hoffmann, S. G. Rogers, R. Fraley, and R. N. Beachy. 1986. Delay of disease development in transgenic plants that express the tobacco mosaic virus coat protein gene. Science 232:738–743.

Price, P. W., C. E. Bouton, P. Gross, B. A. M. Pheron, J. N. Thompson, and A. E. Weiss. 1980. Interactions among three trophic levels: Influence of plants on interactions between insect herbivores and natural enemies. Annu. Rev. Ecol. Syst. 11:41–65.

Prokopy, R. J. 1993. Stepwise progress toward IPM sustainable agriculture. IPM Practitioner 15:1–4.

Putter, C. A. J., and N. A. Van der Graaff. 1989. Information needs in plant protection. Pp. 3–11 in Crop protection information: An international perspective, K. M. Harris and P. R. Scott, eds. Wallingford, U.K.: CAB International.

Rabb, R. L., and F. E. Guthrie, eds. 1972. Concepts of Pest Management. Raleigh: North Carolina State University.

Radosevich, S. R., and J. S. Holt. 1984. Weed Ecology: Implications for Vegetation Management. New York: Wiley & Sons.

Rainwater, C. F. 1962. Where we stand on boll weevil control and research. Pp. 10–19 in Proceedings of the Boll Weevil Research Symposium. Mississippi State: Mississippi State College.

Reichelderfer, K. H. 1979. Economic Feasibility of a bc Technology: Using a Parasitic Wasp, *Pediobius foveolatus*, to Manage Mexican Bean Beetle on Soybeans. Agricultural Economic Report No. 430. Washington, D.C.: U.S. Department of Agriculture.

Reichelderfer, K. H. 1981. Economic feasibility of biological control of pests. Pp. 403–417 in Biological Control in Crop Production, G. C. Papvizes et al., eds. New York: Allanheld, Osmun.

Reichelderfer, K. 1985. Factors affecting the economic feasibility of the biological control of weeds. Pages 135–144 in Proceedings of the VI International Symposium on Biological Control of Weeds, 19–25 August 1984, Vancouver, Canada, E. S. Delfosse, ed. Ottawa: Agriculture Canada.

Reis, R. P., G. R. Noel, and E. R. Swanson. 1983. Economic analysis of alternative control methods for soybean cyst nematode in southern Illinois. Plant Dis. 67:480–483.

Ridings, W. H., D. J. Mitchell, C. L. Schoulties, and N. E. El-Gholl. 1978. Biological control of milkweed vine in Florida citrus groves with a pathotype of *Phytophthora citrophthora*. Pp. 224–240 in Proceedings of the IV International Symposium on Biological Control of Weeds, T. E. Freeman, ed. Gainesville, Fla.: University of Florida.

Riggs, R. D., and J. A. Wrather, eds. 1992. Biology and Management of the Soybean Cyst Nematode. St. Paul, Minn.: American Phytopathological Press.

Rishbeth, J. 1963. Stump protection against fomesannosus. III: Inoculation with *Peniophora gigantea*. Ann. Appl. Biol. 52:63–77.

Roach, S. H., J. E. DuRant, and M. E. Roof. 1990. Cotton performance after boll weevil eradication when left untreated, sprayed with insecticides at least weekly, and sprayed as needed, based on scouting reports: Second year results. Pp. 275–277 in Proceedings of the 1990 Beltwide Cotton Conference. Memphis, Tenn.: National Cotton Council of America.

Robb, R. L., and J. R. Bradley. 1968. The influence of host plants on parasitism of eggs of the tobacco hornworm. J. Econ. Entomol. 61:1249–1252.

Rook, S. P., and G. A. Carlson. 1985. Participation in pest management groups. Am. J. Agric. Econ. 67:563–566.

Rowe, R. C. 1994. Status of the national late blight epidemic and tips for management in 1994. Resistant Pest Manag. Newslet. (Spring):2–3.

Ryan, C. A. 1983. Insect-induced chemical signals regulating natural plant protection responses in variable plants and herbivores. Pp. 43–60 in Natural and Managed Systems, R. F. Denno and M. S. McClure, eds. Orlando, Fla.: Academic Press.

Sands, D. C., E. J. Ford, and R. V. Miller. 1990. Manipulation of broad host-range fungi for biological control of weeds. Weed Technol. 4:471–474.

Schroth, M. N., J. G. Hancock, and A. R. Weinhold. 1992. Biological approaches to the control of plant diseases. Pp. 102–122 in Beyond Pesticides: Biological Approaches to Pest Management in California, T. Beall, ed. Oakland, Calif.: University of California.

Schuster, M. F., and M. Calderon. 1986. Interactions of host plant resistant genotypes and beneficial insects on cotton ecosystems. Pp. 84–97 in Interactions of Plant Resistance and Parasitoids and Predators of Insects, R. D. Boethel and R. D. Eikenbary, eds. New York: Wiley & Sons.

Schwalbe, C. P. 1993. Biologically based regulatory pest management. Pp. 404–409 in Pest Management: Biologically Based Technologies, R. D. Lumsden and J. L. Vaughn, eds. Washington, D.C.: American Chemical Society.

Scott, P. R., and K. M. Harris. 1989. Information dissemination techniques and their importance in crop protection. Pp. 13–26 in K. M. Harris and P.R. Scott, eds. Crop Protection Information: An International Perspective. Wallingford, U.K.: CAB International.

Shennan, C., L. E. Drinkwater, A. H. C. van Bruggen, D. K. Letourneau, and F. Workneh. 1991. Comparative study of organic and conventional tomato production systems: An approach to on-farm systems studies. Pp. 109–144 in Sustainable Agriculture Research and Education in the Field—A Proceedings. Washington, D.C.: National Academy Press.

Simmonds, N. W., ed. 1979. Evolution of Crop Plants. New York: Longman.

Smith, G. L., T. C. Cleveland, and J. C. Clark. 1964. Cost of cotton insect control with insecticides at Tallulah, La. Agricultural Research Service Publ. No. 33–96. Washington, D.C.: Agricultural Research Service, U.S. Department of Agriculture.

Smith, R. F. 1969. The new and the old in pest control. Pp. 21–30 in Proceedings of the Accadimi Nazionale dei Lincei. Rome: Accadimi Nazionale dei Lincei.

Smith, R. F., and R. van den Bosch. 1967. Integrated control. Pp. 295–340 in Pest Control: Biological, Physical, and Selected Chemical Methods, W. W. Kilgore and R. L. Doutt, eds. New York: Academic Press.

Spellig, T., M. Bölker, F. Lottspeich, R. W. Frank, and R. Kahmann. 1994. Pheromones trigger filamentous growth in *Ustilago maydis*. EMBO J. 13:1620–1627.

Staskawicz, B. J., D. Dahlbeck, and N. T. Keen. 1984. Cloned avirulence gene of *Pseudomonas syringae* pv. glycinea determines race-specific incompatibility on *Glycine max* (L.). Merr. Proc. Natl. Acad. Sci. USA 81:6024–6028.

Stern, V. M., R. F. Smith, R. van den Bosch, and K. S. Hagen. 1959. The integrated control concept. Hilgardia 29:81–101.

Stirling, G. R. 1991. Biological Control of Plant Parasitic Nematodes. Wallingford, Oxon, U.K.: CAB International.

Suber, E. F., and J. W. Todd, eds. 1980. Summary of Economic Losses Due to Insect Damage and Costs of Control in Georgia, 1971–1976. Pub. No. 7:5–7. Athens, Ga.: Agricultural Experiment Station, University of Georgia.

Suneson, C. A., K. O. Rachie, and G. S. Khush. 1969. A dynamic population of weedy rye. Crop Sci. 9:121–124.

Supkoff, D. M., D. B. Joley, and J. J. Marois. 1988. Effect of introduced biological control organisms on the density of *Chondrilla juncea* in California. J. Appl. Ecol. 25:1089–1095.

Sutton, D. L. 1977. Grass carp *Ctenopharyngodon idella* Val. in North America. Aquatic Bot. 3:157–164.

Sutton, D. L., and V. V. Vandiver. 1986. Grass Carp: A Fish for Biological Management of Hydrilla and Other Aquatic Weeds in Florida. Bulletin 867. Gainesville, Fla.: Florida Agricultural Experiment Station, University of Florida.

TeBeest, D. O. 1988. Additions to host range of *Colletotrichum gloeosporioides* f. sp. *aeschynomene*. Plant Dis. 72:16–18.

TeBeest, D. O., and G. E. Templeton. 1985. Mycoherbicides: Progress in the biological control of weeds. Plant Dis. 69:6–10.

Tette, J. P., and B. J. Jacobsen. 1992. Biologically intensive pest management in the tree fruit system. Pp. 83–106 in Food, Crop Pests, and the Environment: The Need and Potential for Biologically Intensive Integrated Pest Management, F. G. Zalom and W. E. Fry, eds. St. Paul, Minn.: APS Press.

Thurman, E. M., D. A. Goolsby, M. T. Meyer, and D. W. Kolpin. 1991. Herbicides in surface waters of the midwestern United States: The effect of spring flush. Environ. Sci. Technol. 25:1794–1796.

Thurston, H. D. 1990. Plant disease management practices of traditional farmers. Plant Dis. 74:96–102.

Tiedje, J. M., R. K. Colwell, Y. L. Grossman, R. E. Hodson, R. E. Lenski, R. N. Mack, and P. J. Regal. 1989. The release of genetically engineered organisms: A perspective from the Ecological Society of America. Ecology 70:298–315.

Tisdell, C. A., B. A. Auld, and K. M. Menz. 1984. On assessing the value of biological control of weeds. Protect. Ecol. 6:169–79.

Trujillo, E. E., M. Aragaki, and R. A. Shoemaker. 1988. Infection, disease development, and axenic cultures of *Entyloma compositarum*, the cause of hamakua pamakani blight in Hawaii. Plant Dis. 72:355–357.

Tucker, J. M., and J. D. Sauer. 1958. Aberrant Amaranthus populations of Sacramento-San Joaquin delta, California. Madrono 14:252–261.

Tumlinson, J. H., W. J. Lewis, and L. E. M. Vet. 1993. How parasitic wasps find their hosts. Sci. Am. 268:100–106.

Turner, C. E. 1992. Biological approaches to weed management. Pp. 32–67 in Beyond Pesticides: Biological Approaches to Pest Management in California, T. Beall, ed. Oakland, Calif.: University of California.

U.N. Food and Agriculture Organization. 1967. Report of the First Session of the FAO Panel of Experts on Integrated Pest Control. Rome: U.N. Food and Agriculture Organization.

U.S. Department of Agriculture, Agricultural Biotechnology Research Advisory Committee. 1991. Guidelines for Research Involving Planned Introduction into the Environment of Genetically Modified Organisms. Document No. 91–04. Washington, D.C.: U.S. Department of Agriculture, Office of Agricultural Biotechnology.

U.S. Department of Agriculture, Agricultural Research Service. 1993. Alternatives to Methyl Bromide: Assessment of Research Needs and Priorities, E. L. Civerolo, S. K. Narange, R. Ross, K. W. Vick, and L. Greczy, eds. Washington, D.C.: U.S. Department of Agriculture, Office of Agricultural Biotechnology.

U.S. Department of Agriculture, Agricultural Research Service. 1994. Leafy spurge is united with old enemy. Agric. Res. 42:20–22.

U.S. Department of Agriculture, Agricultural Research Service. 1995. Tackling Wheat Take-All. Agric. Res. 43:4–7.

U.S. Environmental Protection Agency. 1987. Agricultural Chemicals in Ground Water: Proposed Pesticides Strategy. Washington, D.C.: U.S. Environmental Protection Agency.

U.S. Environmental Protection Agency. 1988. Pesticides in Ground Water Data Base: Interim Report. Washington, D.C.: U.S. Environmental Protection Agency.

U.S. Environmental Protection Agency. 1990. National Pesticide Survey: Phase 1 Report. Report No. PB91–125765. Springfield, Va.: U.S. Department of Commerce, National Technical Information Service.

U.S. Environmental Protection Agency. 1994a. Microbial Pesticides: Experimental Use Permits and Notifications, Final Rule. Fed. Reg. 59(169):5600–45615.

U.S. Environmental Protection Agency. 1994b. Plant-Pesticides Subject to the Federal Insecticide, Fungicide, and Rodenticide Act and Federal Food Drug and Cosmetic Act: Proposed Policy. Fed. Reg. 59(225):60496–60547.

U.S. Office of Technology Assessment. 1990. Beneath the Bottom Line: Agricultural Approaches to Reduce Agrichemical Contamination of Groundwater. Washington, D.C.: U.S. Government Printing Office.

U.S. Office of Technology Assessment. 1993. Harmful Non-Indigenous Species in the United States, OTA-F-565. Washington, D.C.: U.S. Government Printing Office.

Van Driesche, R. G. 1991. Building biological control institutions for the twenty-first century. Pp. 119–124 in Proceedings of the Workshop on Biological Control of Pests in Canada, A. S. McClay, ed. Vegreville, Alberta, Canada: Alberta Environmental Centre.

van Lenteren, J. C. 1989. Implementation and commercialization of biological control in West Europe. International Symposium on Biological Control Implementation, Proceedings and Abstracts. N. Am. Plant Protect. Org. Bull. No. 6:50–70.

Wade, P. M. 1990. Physical control of aquatic weeds. Pp. 93–135 in Aquatic Weeds: The Ecology and Management of Nuisance Aquatic Vegetation, A. H. Pieterse and K. J. Murphy, eds. New York: Oxford University Press.

Wapshere, A. J. 1982. Biological control of weeds. Pp. 47–56 in Biology and Ecology of Weeds, W. Holzner and N. Numata, eds. The Hague: W. Junk Publishers.

Watson, A. K. 1985. Host specificity of plant pathogens in biological weed control. Pp. 557–586 in Proceedings of the VI International Symposium on Biological Control of Weeds, E. S. Delfosse, ed. Ottawa: Agriculture Canada.

Weller, D. M., and L. S. Thomashow. 1993. Microbial metabolites with biological activity against plant pathogens. Pp. 173–180 in Pest Management: Biologically Based Technologies, R. D. Lumsden and J. L. Vaughn, eds. Washington, D.C.: American Chemical Society.

Whitham, S., S. P. Dinesh-Kumar, D. Choi, R. Hehl, C. Corr, and B. Baker. 1994. The product of the tobacco mosaic virus resistance gene N: Similarity to Toll and the interleukin-1 receptor. Cell 78:1101–1115.

Whitten, M. J., and J. G. Oakeshott. 1990. Biocontrol of insects and weeds. Pp. 123–142 in Agricultural Biotechnology, G.J. Persley, ed. Cambridge, U.K.: University Press.

Wilson, C. L., and W. J. Janisiewicz. 1995. EPA okays biofungicides from ARS research. Agric. Res. 43:23.

Wrather, J. A., and G. L. Sciumbato. 1995. Soybean disease loss estimates for the southern United States during 1992 and 1993. Plant Dis. 79:84–85.

Wrather, J. A., A. Y. Chambers, J. A. Fox, W. F. Moore, and G. L. Sciumbato. 1995. Soybean disease loss estimates for the southern United States, 1974–1994. Plant Dis. 79:1076–1079.

Zalom, F. G., and W. E. Fry. 1992. Biologically intensive IPM for vegetable crops. Pp.
    107–165 in Food, Crop Pests, and the Environment: The Need and Potential for
    Biologically Intensive Integrated Pest Management, F. G. Zalom and W. E. Fry, eds.
    St. Paul, Minn.: APS Press.
Zalom, F. G., R. E. Ford, R. E. Frisbie, C. R. Edwards, and J. P. Tette. 1992. Integrated
    pest management: Addressing the economic and environmental issues of contempo-
    rary agriculture. Pp. 1–12 in Food, Crop Pests, and the Environment: The Need and
    Potential for Biologically Intensive Integrated Pest Management, F. G. Zalom and
    W. E. Fry, eds. St. Paul, Minn.: APS Press.

About the Authors

Ralph W. F. Hardy, *Chair*, is the former president and chief executive officer of Boyce Thompson Institute for Plant Research, Inc., Ithaca, N.Y. He received his Ph.D. degree in biochemistry from the University of Wisconsin-Madison. He co-founded the National Agricultural Biotechnology Council and will become its president in 1996. His broad research interests encompass biological nitrogen fixation and photosynthesis, and biotechnology. Hardy has served the National Research Council as a member of the Board on Agriculture, Board on Biology, Commission on Life Sciences, and Board on Science and Technology for International Development.

Roger N. Beachy is holder of the Scripps Family Chair, is head of the Division of Plant Biology, and is a member of the Department of Cell Biology at The Scripps Research Institute, La Jolla, California. He is also co-director of the International Laboratory for Tropical Agricultural Biotechnology. Beachy received his Ph.D. degree in plant pathology from Michigan State University and did post-doctoral research at the University of Arizona in Tucson and at Cornell University. His research interests include plant virology and phytopathology, plant gene expression, and agricultural biotechnology.

Harold Browning, an associate professor of entomology at the University of Florida's Citrus Research and Education Center, earned his Ph.D. degree from the University of California at Riverside, where he also conducted post-doctoral research in the Department of Entomology. His research program focuses on relationships between arthropod pests and their natural enemies and the manipulation of indigenous and exotic parasites and predators to enhance natural control processes in perennial subtropical and tropical production systems.

Jerry D. Caulder is currently chairman, president, and chief executive officer of Mycogen Corporation, San Diego, California. Caulder concurrently serves on the boards of directors of the Biotechnology Industry Organization; Environmental Services & Engineering, Inc.; and Applied Genetics. He is also a member of the Advisory Council on Small Business and Agriculture of the Federal Reserve Bank of San Francisco. Caulder earned M.S. and Ph.D. degrees in agronomy and plant physiology at the University of Missouri. He retains an ownership interest in a cotton farm in Missouri.

Raghavan Charudattan is professor of plant pathology at the University of Florida. He received his Ph.D. degree in plant pathology and mycology from the University of Madras, India, and was a post-doctoral research plant pathologist at the University of California at Davis. His research involves biological and integrated controls of weeds using plant pathogens as classical and bioherbicide agents, host-pathogen interactions, diseases of aquatic plants, fungal toxins, and epidemiology. He is a founding editor of the journal *Biological Control: Theory and Application in Pest Management.*

Peter Faulkner, a Career Investigator of the Medical Research Council of Canada and a professor of microbiology at Queen's University, Kingston, Ontario, Canada, received his Ph.D. degree in neurobiochemistry from McGill University, Montreal. As a career study, however, he chose the biochemistry and molecular genetics of insect viruses. Most recently his work has focused on studying the early stages of viral interactions with pest insects and developing strategies to construct recombinant baculoviurses that would be acceptable for release as a class of pest-control agents that would present a minimal intrusive effect on the natural ecosystem.

Fred L. Gould earned his Ph.D. degree in ecology and evolution at the State University of New York at Stony Brook. He is currently a professor of entomology at North Carolina State University. His research interests include ecological genetics of pest adaptation to chemical, biological, and cultural control tactics. His major emphasis in recent years has been focused on developing strategies for sustainable use of transgenic crops that produce insecticidal proteins derived from the bacterium *Bacillus thuringiensis.*

Maureen Kuwano Hinkle, director of agricultural policy for the National Audubon Society, received her B.A. degree in political science from Wellesley College. Hinkle is primarily involved in policy analysis and technology assessment concerning legislation and implementation of farm bills, pesticide regulation, wetlands conservation, and associated issues. Her research interests include integrated pest management, conservation tillage, exotic species, and uses of technology in pest management.

Bruce A. Jaffee is an associate professor in the Department of Nematology at the University of California at Davis. He received his M.S. and Ph.D. degrees in plant pathology from Cornell University. His research concerns the biological control of plant-parasitic nematodes and the population biology of soil-inhabiting nematodes and their natural enemies. His most recent work has focused on suppression of plant-parasitic nematodes by resident or introduced fungi.

Mary K. Knudson is an assistant research scientist in the School of Public Health, Department of Population and International Health at the University of Michigan. She earned her M.S. degree in plant breeding-plant genetics from the University of Wisconsin-Madison and her Ph.D. degree in agricultural and applied economics from the University of Minnesota-Twin Cities. Knudson has been involved in research concerning such issues as public regulation of agricultural biotechnology field tests, the impact of intellectual property rights on genetic diversity, and agricultural diversity to meet future needs.

W. Joe Lewis is a research entomologist with the Insect Biology and Population Management Research Laboratory, Agricultural Research Service, U.S. Department of Agriculture. He is also a founding editor of the journal *Biological Control: Theory and Application in Pest Management.* His research includes foraging behavior of beneficial insects; multitrophic level interactions between parasitoids, herbivores, and plants, with an emphasis on the role of semiochemicals; and the development of an ecologically based pest management system for cotton and other row crops. He earned his M.S. degree and Ph.D. degree in entomology at Mississippi State University.

Joyce E. Loper is a research plant pathologist with the Agricultural Research Service, U.S. Department of Agriculture, and an associate professor of plant pathology at Oregon State University. She received her M.S. degree in plant pathology from the University of California at Davis and Ph.D. degree in plant pathology from the University of California at Berkeley. Her research work includes the molecular genetics, ecology, and physiology of bacteria that inhabit the plant rhizosphere and suppress plant disease caused by soil-borne pathogens. She directs a laboratory group studying biological control of plant diseases, focusing on elucidation of mechanisms involved in biological control.

Daniel L. Mahr earned his Ph.D. degree in entomology at the University of California at Riverside. He is professor of entomology at the University of Wisconsin-Madison and entomologist with the university's extension service. He conducts research and extension programs to develop improved pest management practices and is responsible for coordinating extension biological control programs for the university's Department of Entomology. He is concurrently the project director of the newsletter *Midwest Biological Control News* and coordinator of a series of pest management manuals focusing on biological control practices in the north-central United States.

Neal K. Van Alfen is head of the Department of Plant Pathology and Microbiology at Texas A&M University. He earned his M.S. degree from Brigham Young University and his Ph.D. degree in plant pathology from the University of California at Davis. His primary research has been on the mechanisms of virulence expression by plant pathogens and methods of reducing pathogen virulence for biological control of plant diseases. Current research interests include biological control of forest diseases, viruses of fungi, and effects of disease on plant/water relations.

Index

A

Adaptation, 75
Agency for International Development, 93
Agribusiness, 61-62
Agrobacterium radiobacter, 107
Aldrin, 23
Alligator weed, 13, 37
Antibiotics, 57, 78
Apples, 24
Aquatic weeds, 13, 36-37
Arthropod management
  appetite suppression, 34
  behavioral strategies, 77
  biological-control organisms for, 46
  chemical signaling strategies, 77-78
  in cotton production, 30-31
  cultural techniques for, 17-21
  exotic pests, 32
  genetic engineering strategies, 79-81
  growth of chemical insecticide use, 23, 24
  historical biological strategies, 12-13
  host-range predictions, 102
  host selection/specificity dynamics, 81
  insecticide selectivity, 48
  in IPM, 25-26
  nontarget effects, 103
  pest resistance problems, 26-28
  problems created by pesticide use, 29
  use of disease pathogens for, 79
Arthropods, as biological-control
    organisms
  aquatic weed management, 37
  characteristics, 46
Australian ladybird beetle, 13

B

Bacillus thuringiensis, 47, 64, 76, 78-79, 98-99, 103, 113
Baculoviruses, 79
Banana plants, 17
Benomyl, 29
Biodiversity, 82-84
Biological-control organisms
  aquatic weed management, 36-37
  in citrus farming, 50-51
  for managing plant viruses, 34
  cover crop design, 45
  cultural practices to encourage, 20-21
  current registration, 96

definition, 46
early farm practices, 12-17
EBPM principles, 46-47
ecosystem interaction, 43-44, 76-82
environmental persistence, 110-111
experimental demonstrations, 18-19
in IPM, 25-26
microbial, 84-85, 103
molecular mechanisms, 76
natural reservoirs, 82-83
nontarget effects, 74, 100-105
objectives, 46
pathogenic potential, 105-108
regulatory environment, 112-114
research needs, 5, 76-82
risk assessment, 8-9, 97, 100, 105, 108-109
scale of use, 110
self-perpetuation of, 110-111
for soybean cyst nematode, 33
specificity, 81
success rate, 70
supply system, 55, 85-86
use of disease pathogens as, 79
Biological-control products
definition, 47
potential risks, 98-100
predator-prey population dynamics, 102-103
small-market support, 64, 97-98
specificity, 47, 71
Black scale, 65
Boll weevil, 30-31
Botrytis rot of cyclamen, 29
Broad-spectrum pesticides, 1, 26
naturally occurring, 47
obstacles to continued use, 26-40
use in cotton farming, 30-31
Brown soft scale, 29

C

California red scale, 18, 50, 65
Cancer, 40-41
Certification, 64
Chemical signaling, 77-78
Chestnut blight, 17, 80

Chlordane, 23, 38
Citrus farming
biological pest control, 50-51
early biological control mechanisms, 12, 13
pest-management cooperatives, 63, 65
pesticide-associated problems, 29
pesticide use, 24
Citrus red mite, 18, 29
Citrus rust mite, 50
Collego^R, 104, 106
Commercial development
biological management of diseases, 17
genetically engineered plants, 23
growth of chemical pesticide industry, 23-25
health risks for workers, 98
pesticide industry trends, 28-29
regulatory obstacles, 48, 112-115
small-market products, 64, 97-98
Communication among organisms, 77-78
Consultants, 7-8, 10, 62
Corn
growth of pesticide use, 24
hybridization, 21
leaf blight, 21
Cost-benefit analysis
data base for, 90-91
feasibility of EBPM, 49-56
measuring direct and indirect effects in, 89-90
role of, 89
Cotton, 24
arthropod management strategies, 30-31
pesticide use trends, 24
Cottony cushion scale, 13, 18, 29, 51
Cover crops, 20, 45
Crop rotation, 18-19, 20
for soybean cyst nematode management, 33
Cropping patterns
for managing plant viruses, 34
historical applications for biological control, 17-21
landscape ecology, 74-75
predictive models for, 86-87
row crop pests, 35

use of cover crops, 45
wheat monoculture for disease control, 22
Crown gall, 17, 107
Cultural approaches, 2, 17
Cryphonectria parasitica, 79, 80
Cyst nematode, 18-19, 33

D

DDT, 23, 26, 28, 29, 30, 38, 39
Dibromochloropropane, 23, 33
Dichloropropenes, 23
Dieldrin, 23
Disease management
  biological-control organisms for, 46-47, 51
  development of resistant cultivars for, 82
  early biological strategies, 17
  fungicide use as obstacle to, 29
  genetic diversity of crop species and, 21
  genetic engineering for, 22-23
  host selection dynamics, 81
  limitations of chemical strategies, 34-35
  plant resistance for, 47-49
  resistance in pathogens, 28
  soil-borne diseases, 32-34
  take-all disease in wheat, 22
  through monoculture planting, 22
  use of disease pathogens for, 79
Durability
  EBPM objectives, 3, 4, 42, 43, 115
  monitoring for pest resistance, 64-68
  of plant-host resistance, 87-88

E

EBPM. *See* Ecologically based pest management
Ecologically based pest management (EBPM)
  biodiversity and, 82-84
  continuum of tactics, 108
  economic feasibility, 49-56, 90
  environmental risks, 100

integrated pest management and, 3, 10, 86, 94
  knowledge base, 3, 43, 69, 70-71
  leadership, 93-94
  objectives, 3-4, 42-44, 115
  principles, 2-3, 10, 44-46
  research needs, 4-8
  use of biological-control organisms, 46-47
  use of resistant plants, 47-49
  use of synthetic chemicals, 47
  See also Implementation of EBPM
Economic feasibility of EBPM, 49-56
Ecosystem functioning
  cover crop planning and, 45
  feedback, 43
  gene transfers between microorganisms, 105-108
  integration of EBPM, 43-46
  landscape ecology, 74-75
  managed systems and, 72-76, 86-87
  microbial communities, 72, 73, 84-85, 103
  potential risks of EBPM, 100
  predator-prey population dynamics, 72-76, 102-103
  research needs, 5, 6-7
  stability, 72
Education and training
  for EBPM implementation, 7-8
  introducing new technologies/practices, 60-62
  pest-management consultants, 62
  role of the university, 62-63
Environmental Protection Agency, 93, 96, 98, 109
  creation of, 38-39
  recommendations for, 10, 114
  regulation of biological-control organisms, 110, 112, 113
  regulatory suspension of chemical pesticides, 38-39
Ethylene dibromide, 23, 33
Eurasian water milfoil, 13
Exotic species, 32
  as biological-control organisms, 101

federal regulations for managing, 112-113
weeds, 35
Extension system, 7-8, 60-61, 93

F

Farming practice
accessibility of research, 85
chemical runoff, 39-40
cultural techniques, historical
development of, 17-21
early biological management strategies,
12-17
groundwater contamination, 40
grower cooperatives, 63-64, 65
information flow, 58-59
operational models, 60
pest management knowledge needs, 59-60
risk behavior, 54-55
whole-farm system, 44
Federal Insecticide, Fungicide, and
Rodenticide Act, 37, 38, 112-113
Federal Plant Pest Act, 112-113
Fire blight, 57
Food and Drug Administration, 112
Food contamination
by biological-control organisms, 98-99
pesticide tolerances, 26
Forest management, 17
Frost damage, 57, 73
Fungicides, biological, for aquatic weed
management, 37
Fungicides, chemical
disease problems created by, 29
historical development, 23
limitations of, 32-34
pest resistance to, 28
Fusarium wilt, 17, 32

G

Gene transfer in nature, 105-108
Genetic engineering
to affect host selection/specificity, 81
for appetite suppression in pests, 35
of arthropod predators, 79-81

commercial development, 23
concerns about, 74
durability of resistance genes, 87-88
expression of coat protein genes, 22-23
hybrid plant breeding, 21
for plant resistance, 49, 82
potential human health risks, 99-100
risk assessment in, 97
toxin-encoding, risks of, 103
transgenic plant breeding, 21-22
Genetic uniformity, 21
Geographic information systems, 7, 75
Grass carp, 36
Grazing land, 35
Green manure, 20, 73
Groundwater contamination, 40
Grower cooperatives, 63-64, 65

H

Herbicides, biological
for aquatic weed management, 36-37
obstacles to development, 37
for weed management, 104
Herbicides, chemical
for aquatic weed control, 36
historical development, 23, 24
plant resistance to, 28
runoff problems, 39-40
Historical developments
arthropod management, biological
strategies for, 12-13
boll weevil control strategies, 30-31
chemical pesticides, 23-25, 69
cultural practices for biological control,
17-21
disease management, biological
strategies for, 17
genetically engineered plants, 22-23
origins of weeds, 106
plant breeding, 21
public concerns about chemical
pesticides, 37-40
weed management, biological strategies
for, 13
Human health
acute effects of pesticide exposure, 40

chronic effects of pesticide exposure, 40-41
genetic engineering, potential risks in, 99-100
potential effects of biological-control products, 98-100
potential effects of resistant cultivars, 99
risk assessment of biological controls, 97, 111-112
scale of use of biological-control organisms and, 110
use of chemical pesticides and, 26, 37-40
Hybridization, 21
gene transfer between crops and weeds, 105-108
weed-eating fish, 36
Hydrilla, 13, 36

I

Implementation of EBPM
certification, 64
collective action for, 63
corporate-level, 66-67
demonstration projects, 60-61
economic considerations, 49-56
grower cooperatives in, 63-64, 65
impact of new technologies, 89, 108-109
information transfer, 7-8, 56-63
initial targets, 56
ongoing monitoring activities, 64-68
oversight activities, 8-10
planning for, 41
requirements for success, 10
research funding for, 94-95
research needs, 7, 69-71, 84-86, 88
risk aversion and, 54-56
small-market support for, 64, 97-98
socioeconomic issues, 89-91
supply of resources for, 85-86
Indole acetic acid, 23
Information management
for cost-benefit analysis, 90-91
EBPM guidelines, 114-115

for implementation of EBPM, 7-8, 58-63
knowledge base for risk assessment, 108-109
participants in, 58-59, 63
patterns of information flow, 58-60
private sector role, 61-62
Insecticides, chemical
historical development, 24
regulatory suspension, 38
selectivity, 48
Integrated pest management (IPM)
corporate application, 66-67
EBPM and, 3, 10, 86, 94
implementation, 25
information flow, 56-58
interdisciplinary relations in, 92
objectives, 3
theoretical basis, 25
Intercropping, 20
IPM. *See* Integrated pest management
IR-4 Program, 98

J

Japanese beetle, 29
Johnsongrass, 106

K

Klamath weed, 13

L

Laboratory testing, 102, 103
Landscape ecology, 74-75
Leadership issues, 93-94
Livestock management, 35

M

Malathion, 28, 30-31
Manure, 18-19
Methyl bromide, 23, 35
Monitoring pest behavior
to evaluate economic feasibility of EBPM, 52-53

grower-friendly systems, 85
interactions among organisms, 76-82
landscape studies, 74-75
pathogen host range, 101-102
for pest management strategies, 77
population dynamics, 73-74, 75-76,
   102-103
research needs, 6-7
signaling mechanisms, 77-78
for signs of resistance, 64-68
Monoculture planting, 22

N

National Environmental Protection Act,
   113
National Institutes of Health, 93
National Science Foundation, 93

O

Operational models
farm practice, 60
pathogen in microcosm, 102

P

Parathion, 29
Paris green, 12
Pesticides, biological
historical use of, 12-13
risk assessment, 97
Pesticides, chemical
chronic exposure, 40-41
in cotton production, 30-31
in EBPM, 47
groundwater contamination, 40
historical development, 11, 23-25, 69
human health concerns, 37-41
in IPM, 25
limitations of, 29-37
problems associated with, 1, 11-12, 26-
   29, 100
regulatory suspension, 38-39
research and development trends, 28-29
risk assessment, 97
for row crop application, 35

selectivity, 48
usage trends, 24-25
Pheremones, 78
Phlebia gigantea, 17
Phyllosphere, 72-73
Plant breeding, 21-23
implementation of EBPM, 54
for plant resistance, 49
for viral resistance, 34-35
Population dynamics, 73-74, 75-76, 102-
   103
Potatoes, 24
Primicarb, 48
Professional societies, 93
Profitability
cost of exotic pest invasions, 32
cost of pesticide resistance, 27-28
cost of plant virus damage, 34
cost of soil-borne diseases, 32, 33
EBPM objectives, 3, 4, 42, 43
economic feasibility of EBPM, 49-52,
   90
market factors, 53-54
pest-control factors, 52-53
risk factors, 54-56
Propham, 23
Public intervention/oversight
aquatic weed control, 36
coordination of government groups, 93-
   94, 113-114
current limitations, 112-114
development of new biological
   products, 64
EBPM, 8-10
guidelines for, 10, 114-115
human health concerns, 37-41, 98-100
for information transfer, 7-8, 62
introduction of new products, 108-109
knowledge base for, 9, 108-109
microbial herbicides, 37
need for, 8, 96
opportunities for improving, 113-115
pesticide tolerances in foodstuffs, 26
priority areas, 109-111
private economic interest and, 55
regulatory obstacles to pesticide
   development, 48

regulatory suspension of chemical
    pesticides, 38-39
risk assessment, 8-9, 96-97
Pyrethroids, 24
Pyrethrum, 47

R

Rangelands, 35
Remote sensing technology, 75
Research
    accessibility to grower, 85
    demonstration projects, 60-61
    for EBPM implementation, 7, 69-71, 88
    for EBPM success, 4-8
    on ecology of managed systems, 72-76
    on ecosystem interactions, 5, 91
    federal efforts, 93-94
    funding patterns, 92
    health risks for researchers, 98
    on impact of new technologies, 89
    infrastructure for, 94
    institutional approaches for cooperation,
        91
    interactions among organisms, 76-82
    methodological enhancements, 6-7, 84-
        86
    microbial communities, 84-85, 95, 103
    microcosm studies, 102
    multidisciplinary, 6, 91-94
    natural defense systems, 78-79
    natural resource inventory and
        maintenance, 82-83
    new chemical pesticides, 28-29, 35
    operational models for growers, 60
    organic pesticides, 4
    plant resistance, 49
    predictive models for cropping systems,
        86-87
    priority areas, 6, 71
    recommendations, 4
    resources for, 94-95
    on risk attitudes, 91
    on socioeconomic issues, 89-91
    taxonomic, 83-84
Resistance in pests
    to chemical pesticides, 1

cost of, 27-28
cotton boll weevil experience, 30-31
monitoring for, 64-68
as objection to broad-spectrum
    pesticides, 26-28
planning for, 28
research needs, 75
to viruses, 34-35
Resistant cultivars
    durability of, 87-88
    in EBPM, 4, 47-49
    expression of resistance, 104-105
    genetic engineering for, 22-23
    herbicide-resistant biotypes, 28
    mechanisms, 82
    plant breeding for, 21
    potential harmful effects, 99, 103-
        105
    regulatory environment, 114
    soybean cyst nematode, 33
Rhizosphere, 72-73
Rice, 24, 106
Risk assessment
    activities in, 111-112
    criteria, 96-97
    current federal efforts, 112-114
    in EBPM implementation, 8-9
    economic considerations, 54-56
    environmental effects, 100
    for genetic science, 97
    human health considerations, 98-100
    individual attitudes, 91
    knowledge base for, 9, 108-109
    pathogen host-range, 101-102
    pathogenic potential of biological-
        control organisms, 105-108
    persistence of control organisms, 110-
        111
    priority areas, 109-111
    public role in, 96
    in research and production
        environments, 98
    resistant cultivars, potential harmful
        effects of, 99, 103-105
    risk management and, 111-112
    scale of use issues, 110
    standards for, 114-115

Rotenone, 47
Rust diseases, 47-48

S

Safety, as EBPM objective, 3-4, 42
Silent Spring, 37-38
Soil studies
    cover crop effects, 45
    microbial communities, 72
    pest population dynamics, 72-73
    soil-borne diseases, 32-34, 81
Sorghum, 106
Soybeans, 24, 33

T

Take-all disease, 22
Taxonomic research, 83-84
Technology development, regulatory
    review, 9
Tetraethylpyrophosphate, 23
Tobacco, 12
Tobacco mosaic virus, 22
Toxaphene, 23
Toxic Substances Control Act, 113
Transgenic animals/plants, 21-22

U

Uncertainty, 54-55
U.S. Department of Agriculture, 93, 94,
    109, 112
    recommendations for, 10, 114

U.S. Department of Energy, 93
U.S. Department of the Interior, 93

V

Viruses, plant, 34-35
    as biological-control products, 79

W

Water hyacinth, 13, 36
Weed management
    aquatic weeds, 13, 36-37
    bioherbicide, 104
    biological-control strategies, 51
    cultural practices for, 20
    early biological strategies, 13
    herbicide-resistant biotypes, 28
    landscape ecology for, 74-75
    persistence of control organisms, 111
    in rangelands, 35
    sources of weeds, 106
    use of disease pathogens for, 79
    weed composition shifting and, 19
Wheat, 22, 25, 32
White amur, 36
Whole-farming systems
    EBPM in, 44
    pest management in, 1-2

Z

Zineb, 23

Recent Publications of the Board on Agriculture

Policy and Resources

Colleges of Agriculture at the Land Grant Universities: A Profile (1995), 146 pp, ISBN 0-309-05295-5

Investing in the National Research Initiative: An Update of the Competitive Grants Program in the U.S. Department of Agriculture (1994), 66 pp, ISBN 0-309-05235-1

Rangeland Health: New Methods to Classify, Inventory, and Monitor Rangelands (1994), 180 pp., ISBN 0-309-04879-6

Soil and Water Quality: An Agenda for Agriculture (1993), 516 pp., ISBN 0-309-04933-4

Managing Global Genetic Resources: Agricultural Crop Issues and Policies (1993), 450 pp., ISBN 0-309-04430-8

Pesticides in the Diets of Infants and Children (1993), 408 pp., ISBN 0-309-04875-3

Managing Global Genetic Resources: Livestock (1993), 294 pp., ISBN 0-309-04394-8

Sustainable Agriculture and the Environment in the Humid Tropics (1993), 720 pp., ISBN 0-309-04749-8

Agriculture and the Undergraduate: Proceedings (1992), 296 pp., ISBN 0-309-04682-3

Water Transfers in the West: Efficiency, Equity, and the Environment (1992), 320 pp., ISBN 0-309-04528-2

Managing Global Genetic Resources: Forest Trees (1991), 244 pp., ISBN 0-309-04034-5

Managing Global Genetic Resources: The U.S. National Plant Germplasm System (1991), 198 pp., ISBN 0-309-04390-5

Sustainable Agriculture Research and Education in the Field: A Proceedings (1991), 448 pp., ISBN 0-309-04578-9

Toward Sustainability: A Plan for Collaborative Research on Agriculture and Natural Resource Management (1991), 164 pp., ISBN 0-309-04540-1

Investing in Research: A Proposal to Strengthen the Agricultural, Food, and Environmental System (1989), 156 pp., ISBN 0-309-04127-9

Alternative Agriculture (1989), 464 pp., ISBN 0-309-03985-1

Understanding Agriculture: New Directions for Education (1988), 80 pp., ISBN 0-309-03936-3

Designing Foods: Animal Product Options in the Marketplace (1988), 394 pp., ISBN 0-309-03798-0; ISBN 0-309-03795-6 (pbk)

Agricultural Biotechnology: Strategies for National Competitiveness (1987), 224 pp., ISBN 0-309-03745-X

Regulating Pesticides in Food: The Delaney Paradox (1987), 288 pp., ISBN 0-309-03746-8

Pesticide Resistance: Strategies and Tactics for Management (1986), 480 pp., ISBN 0-309-03627-5

Pesticides and Groundwater Quality: Issues and Problems in Four States (1986), 136 pp., ISBN 0-309-03676-3

Soil Conservation: Assessing the National Resources Inventory, Volume 1 (1986), 134 pp., ISBN 0-309-03649-9; Volume 2 (1986), 314 pp., ISBN 0-309-03675-5

New Directions for Biosciences Research in Agriculture: High-Reward Opportunities (1985), 122 pp., ISBN 0-309-03542-2

Genetic Engineering of Plants: Agricultural Research Opportunities and Policy Concerns (1984), 96 pp., ISBN 0-309-03434-5

Nutrient Requirements of Domestic Animals Series and Related Titles

Nutrient Requirements of Laboratory Animals, Fourth Rev. Ed. (1995), 176 pp., ISBN 0-309-05126-6

Metabolic Modifiers: Effects on the Nutrient Requirements of Food-Producing Animals (1994), 81 pp., ISBN 04997-0

Nutrient Requirements of Poultry, Ninth Revised Edition (1994), 156 pp., ISBN 0-309-04892-3

Nutrient Requirements of Fish (1993), 108 pp., ISBN 0-309-04891-5

Nutrient Requirements of Horses, Fifth Revised Edition (1989), 128 pp., ISBN 0-309-03989-4; diskette included

Nutrient Requirements of Dairy Cattle, Sixth Revised Edition, Update 1989 (1989), 168 pp., ISBN 0-309-03826-X; diskette included

Nutrient Requirements of Swine, Ninth Revised Edition (1988), 96 pp., ISBN 0-309-03779-4

Vitamin Tolerance of Animals (1987), 105 pp., ISBN 0-309-03728-X

Predicting Feed Intake of Food-Producing Animals (1986), 95 pp., ISBN 0-309-03695-X

Nutrient Requirements of Cats, Revised Edition (1986), 87 pp., ISBN 0-309-03682-8

Nutrient Requirements of Dogs, Revised Edition (1985), 79 pp., ISBN 0-309-03496-5

Nutrient Requirements of Sheep, Sixth Revised Edition (1985), 106 pp., ISBN 0-309-03596-1

Nutrient Requirements of Beef Cattle, Sixth Revised Edition (1984), 90 pp., ISBN 0-309-03447-7

Further information, additional titles (prior to 1984), and prices are available from the National Academy Press, 2101 Constitution Avenue, NW, Washington, DC 20418, 202-334-3313 (information only); 800-624-6242 (orders only); 202-334-2451 (fax).